ford Science Research Papers

A bibliography of *ab initio* molecular wave functions

W. G. Richards

T. E. H. Walker

R. K. Hinkley

ARENDON PRESS OXFORD 1971

Oxford University Press, Ely House, London W.1

GLASGOW NEW YORK TORONTO MELBOURNE WELLINGTON

CAPE TOWN SALISBURY IBADAN NAIROBI DAR ES SALAAM LUSAKA

ADDIS ABABA BOMBAY CALCUTTA MADRAS KARACHI LAHORE DACCA

KUALA LUMPUR SINGAPORE HONG KONG TOKYO

Set by E.W.C. Wilkins & Assoc. Ltd.,
and printed in Great Britain by
Lowe and Brydone Ltd, Victoria Road, London

PREFACE

Our group in the Physical Chemistry Laboratory in Oxford are users of molecular wave functions rather than primary producers and for a long time we have felt the need of a comprehensive source book containing a bibliography of *ab initio* wave functions. Discussions with other research workers, not only those who compute wave functions but more particularly with experimentalists, have convinced us that this need is a general one.

All scientists who are interested in molecular problems need to know what has already been published and although there are good compilations of spectroscopic data there are no comprehensive accounts of quantum mechanical calculations.

There are partial bibliographies which we have used to supplement the information of our own card index. Chief among these is the *Compendium of* ab initio *calculations of molecular energies and properties* edited by Krauss (National Bureau of Standards, Technical Note 438). This contains many of the references published up to 1967 and gives more details of some calculations than we do here. Another set of references was circulated privately by Clark and Stewart (LCAO molecular orbital wave functions for systems of four or more electrons), now published, *Quart. Rev.* 24, 95 (1970), and a detailed critical bibliography on first-row diatomic hydrides was published in the *Report of the Laboratory of Molecular Structure and Spectra* (Chicago, 1966) by Cade. We have incorporated all the references from these sources and also systematically searched the literature so that the final bibliography contains considerably more references than any previous compilation.

We have listed those calculations that are commonly called *ab initio*. By this we understand inclusion of all electrons and accurate computation of every integral. All work that has been published up until 31 December 1969 is included. Although we are often aware of its existence, we have deliberately excluded any reference to unpublished work.

Each molecule has been treated separately. An indication of the geometry, type of calculation, and energy is given, followed by a list of computed properties. All numerical quantities are in atomic units. We then give a complete chronological bibliography for the molecule so that there should be no need to chase cross references in other parts of the book. We have attempted to avoid the use of initials and esoteric jargon in the hope that readers can avoid turning to a key for explanation. The few common abbreviations used are listed below.

The order of the molecules is necessarily somewhat arbitrary. We have chosen to list diatomic molecules followed by triatomic and tetratomic and then larger molecules. The diatomic molecules are in the alphabetic order defined by Herzberg (*Spectra of diatomic molecules*, Van Nostrand, New York, 1950). The triatomic and tetratomic molecules also follow his order (*Electronic spectra of polyatomic molecules*, Van Nostrand, New York, 1966); that is to say hydrides are taken first. The larger molecules are grouped in a manner that takes some account of chemical similarities and connections and a complete list is given on p. 00.

Although we have attempted to include all the published work and to avoid errors it would be surprising if we had not ommitted some papers and made some mistakes. Far from being dismayed by criticism or information about work we have not included, we would welcome such help and hope to incorporate it in any revision.

Physical Chemistry Laboratory
South Parks Road, Oxford

W.G. RICHARDS
T.E.H. WALKER
R.K. HINKLEY

CONTENTS

DIATOMIC MOLECULES

TRIATOMIC MOLECULES

ABBREVIATIONS

Min.	minimum
Ext.	extended
M.O.	molecular orbital
V.B.	Valence bond
STO	Slater-type orbital
GTO	Gaussian-type orbital
CI	Configuration interaction
FSGO	Floating spherical gaussian orbitals
√	In column headed 'Wave function' indicates that the wave function is published.

$\underline{AlF}$

State	Internuclear distance	Basis set	Energy	Wave function	Reference
$X\,^1\Sigma^+$	3·45	Ext. STO	−341·4718		1
	3·126	Ext. STO	−341·4832	√	2

Properties	*References*
Dipole moment	1
Quadrupole moment	1
Potential curve	2
Orbital energies	2

REFERENCES

1. M. Yoshimine and A.D. McLean, *Int. J. Quantum Chem.* **1S**, 313(1967).
2. A.D. McLean and M. Yoshimine, Supp. to *I.B.M. Jl Res. Dev.* (1967).

Both these wave functions are very near the Hartree–Fock limit, though at different internuclear distances.

$\underline{AlH}$

State	Internuclear distance	Basis set	Energy	Wave function	Reference
$X\,^1\Sigma^+$	3·114	Ext. STO	−242·4634	√	3

Properties	*References*
Spectroscopic constants	3
Ionization potentials	3
Potential curve	3
Dipole moment	1,4,5
Rydberg levels	2
Charge distribution	5
Orbital energies	3
Magnetic properties	7

REFERENCES

1. P.E. Cade and W.M. Huo, *J. chem. Phys.* **45**, 1063 (1966).
2. F. Grimaldi, A. Lecourt, H. Lefebvre-Brion, and C.M. Moser, *J. molec. Spectrosc.* **20**, 341 (1966).
   Rydberg levels using AlH^+ core and further basis set; virtual orbitals
3. P.E. Cade and W.M. Huo, *J. chem. Phys.* **47**, 649 (1967).
   Very near Hartree–Fock limit.
4. P. Politzer, *Chem. Phys. Lett.* **1**, 227 (1967).
   Discussion of dipole moment.
5. P.E. Cade, R.F.W. Bader, W.H. Henneker, and I. Keaveny, *J. chem. Phys.* **50**, 5313 (1969).
   Charge distribution using wave function of 2.
6. F. Grimaldi, A. Lecourt, and C.M. Moser, *Symp. Faraday Soc.* **2**, 59 (1968).
   Dipole moment.
7. E.A. Laws, R.M. Stevens, and W.N. Lipscomb, *Chem. Phys. Lett.* **4**, 159 (1969).
   Magnetic properties from coupled Hartree–Fock.

AlH^+

State	Internuclear distance	Basis set	Energy	Wave function	Reference
$^2\Sigma^+$	3·027	Ext. STO	−242·1930		1

REFERENCES

1. P.E. Cade and W.M. Huo, *J. chem. Phys.* **47**, 649 (1967).
   Very near Hartree–Fock limit.
2. T.E.H. Walker and W.G. Richards, *Symp. Faraday Soc.* **2**, 64 (1968).
   Calculation of spin–orbit coupling constant of $A^2\Pi$ state.

Ar_2

State	Internuclear distance	Basis set	Energy	Wave function	Reference
$^1\Sigma_g^+$		STO		✓	1

Properties	*References*
Orbital energies	1
Potential curve	1

REFERENCE

1. T.L. Gilbert and A.C. Wahl, *J. chem. Phys.* **47**, 3425 (1967). Single configuration wave functions.

ArF^+

State	Internuclear distance	Basis set	Energy	Wave function	Reference
$^1\Sigma^+$	3·5	Gaussian lobe	–625·5399		1

Properties	*Reference*
Potential curve	1

REFERENCE

1. J.L. Liebman and L.C. Allen, *Chem. Communs.* 1355 (1969).

ArH^+

State	Internuclear distance	Basis set	Energy	Wave function	Reference
$^1\Sigma^+$	3·11	One centre	–524·5752		1

Properties	*Reference*
Force constant	1

REFERENCE

1. K.E. Banyard and A. Sutton, *J. chem. Phys.* **46**, 2143 (1967). One centre with min. basis set.

ArO

State	Internuclear distance	Basis set	Energy	Wave function	Reference
$^1\Sigma^+$	Repulsive	Gaussian lobe			1

REFERENCE

1. J.F. Liebman and L.C. Allen, *Chem. Communs.* 1335 (1969).

B_2

State	Internuclear distance	Basis set	Energy	Wave function	Reference
$^5\Sigma_u^-$	3·0769	Min. STO	–49·063	✓	1
	3·0	Elliptical + CI	–49·14530		3
$^3\Sigma_g^-$	3·0769	Min. STO	–48·996	✓	1
	3·0	Elliptical + CI	–49·14020		3
$^3\Pi_u$	3·0769	Min. STO	–49·002	✓	1
	3·0	Elliptical + CI	–49·11438		3
$^3\Pi_g$	3·0	Elliptical + CI	–49·04384		3
$^1\Sigma_g^+$	3·0	Elliptical + CI	–49·06366		3

Properties	*References*
Potential curves	1, 3
Orbital energies	1, 3
Population analysis	1
Term values	1
Charge density maps	2, 4
Dissociation energy	4
Orbital forces	4

REFERENCES

1. A.A. Padgett and V. Griffing, *J. chem. Phys.* **30**, 1286 (1959).
2. A.C. Wahl, *Science, N.Y.* **151**, 961 (1966).
Charge density map from STO wave functions.
3. C.F. Bender and E.R. Davidson, *J. chem. Phys.* **46**, 3313 (1967).
4. R.F.W. Bader, W.H. Henneker, and P.E. Cade, *J. chem. Phys.* **46**, 3341 (1967).
Wave functions from J.B. Greenshields (unpub.).

BF

State	Internuclear distance	Basis set	Energy	Wave function	Reference
$X\,^1\Sigma^+$	2·385	Min. STO	–123·61550	✓	1

State	Internuclear distance	Basis set	Energy	Wave function	Reference
	2·385	Min. STO	−123·67562	√	3
	2·385	Ext. STO	−124·14038	√	5
	2·354	Ext. STO	−124·1664	√	7
	2·391	Ext. STO	−124·1671	√	13, 18

Properties	*References*
Orbital energies	1, 5, 7, 13
Spectroscopic constants	5, 7
Potential curves	5, 7, 13
Dipole moment	1, 3, 5, 7, 18
Quadrupole moment	5
Magnetic properties	2, 9, 10, 14, 16, 19
Atomic population	4, 12
Excited states	6, 8
Charge density	7
Polarizabilities	11
Localized orbitals	15
Transition probabilities	17

REFERENCES

1. B.J. Ransil, *Rev. mod. Phys.* **32**, 239 (1960).
2. B.J. Ransil, *J. chem. Phys.* **34**, 727 (1961).
   Magnetic shielding using wave function of ref. 1.
3. S. Fraga and B.J. Ransil, *J. chem. Phys.* **36**, 1127 (1962).
   Limited CI also carried out.
4. E. Clementi and H. Clementi, *J. chem. Phys.* **36**, 2824 (1962).
5. R.K. Nesbet, *J. chem. Phys.* **40**, 3619 (1964).
6. H. Lefebvre-Brion, C.M. Moser, and R.K. Nesbet, *J. molec. Spectrosc.* **15**, 211 (1965).
   Term values of excited states.
7. W.M. Huo, *J. chem. Phys.* **43**, 624 (1965).
   Charge density contours given.
8. R.K. Nesbet, *J. chem. Phys.* **43**, 4403 (1965).
   Excited states using wave function from ref. 5.
9. R.A. Hegstrom and W.N. Lipscomb, *J. chem. Phys.* **45**, 2378 (1965).
   Magnetic properties using wave function of ref. 7.
10. W.N. Lipscomb, *Adv. mag. Resonance* **2**, 137 (1966).
    Magnetic constants from perturbed, coupled Hartree–Fock wave function.

11. J.M. O'Hare and R.P. Hurst, *J. chem. Phys.* **46**, 2356 (1967).
Polarizability and hyperpolarizability.

12. E.R. Davidson, *J. chem. Phys.* **46**, 3320 (1967).
Atomic population using wave function of ref. 7.

13. A.D. McLean and M. Yoshimine, Supp. to *I.B.M. Jl Res. Dev.* (1967).
Very near Hartree–Fock limit.

14. J.R. de la Vega, D. Ziobro, and H.F. Hameka, *Physica, 's Grav* **37**, 265 (1967).
Magnetic susceptibility.

15. V. Magnasco and A. Perico, *J. chem. Phys.* **47**, 971 (1967).
Localized orbitals using wave function of ref. 1.

16. J.R. de la Vega and H.F. Hemeka, *J. chem. Phys.* **47**, 1834 (1967).
Rotational magnetic moment and anisotropies in diamagnetic susceptibility.

17. S.R. La Paglia, *Theor. Chim. Acta* **8**, 185 (1967).
Transition probabilities, including CI.

18. M. Yoshimine and A.D. McLean, *Int. J. Quantum Chem.* **1S**, 313 (1967).
Same wave function as 13.

19. R.A. Hegstrom and W.N. Lipscomb, *J. chem. Phys.* **48**, 809 (1968).
Magnetic constants from coupled Hartree–Fock theory.

BH

State	Internuclear distance	Basis set	Energy	Wave function	Reference
$X\,^1\Sigma^+$	2·329	Min. STO	–25·057	✓	3
	2·329	Min. STO	–25·0746	✓	7
	2·329	Min. STO + CI	–25·093	✓	15
		V.B.	–25·105		2
	2·329	V.B. + CI	–25·110	✓	4
	2·305	Ext. STO	–25·13147	✓	25
	2·536	V.B. + CI Gaussian lobe	–25·1456	✓	35
	2·336	Natural orbitals + CI	–25·2621		34

Properties	*References*
Dipole moment	3, 5, 22, 33, 34, 35
Spectroscopic constants	8, 10

Properties	*References*
Population analysis	9, 16
Magnetic properties	12, 19, 26, 27, 28, 29, 31, 34, 35
Rydberg levels	21
Charge densities	24, 34
Potential curves	25, 35
Ionization potential	25
Transition probabilities	30
Quadrupole moment	35

REFERENCES

1. J. Higuchi, *Bull. chem. Soc. Japan,* **25**, 1 (1952).
Simplified V.B.

2. S.F. Boys, G.B. Cook, C.M. Reeves, and I Shavitt, *Nature ,Lond.* **178**, 1207 (1956).
Valence bond.

3. R.C. Sahni, *J. chem. Phys.* **25**, 332 (1956).
Min. STO basis.

4. K. Ohno, *J. phys. Soc. Japan,* **12**, 938 (1957).
13 config. V.B., Results also for several excited states.

5. A.C. Hurley, *Proc. R. Soc.* **A248**, 119 (1958).
Semi-empirical calculation.

6. A.C. Hurley, *Proc. R. Soc.* **A249**, 402 (1959).
Semi-empirical treatment of excited states.

7. B.J. Ransil, *Rev. mod. Phys.* **32**, 245 (1960).
Min. STO.

8. B.J. Ransil, *Rev. mod. Phys.* **32**, 239 (1960).
Scope of diatomic programme.

9. S. Fraga and B.J. Ransil, *J. chem. Phys.* **34**, 727 (1961).
Population analysis.

10. S. Fraga and B.J. Ransil, *J. chem. Phys.* **35**, 669 (1961).
Spectroscopic constants.

11. S. Frage and B.J. Ransil, *J. chem. Phys.* **36**, 1127 (1962).
Limited C.I. on min. basis set function.

12. C.W. Kern and W.N. Lipscomb, *J. chem. Phys.* **37**, 260 (1962).
Magnetic shielding.

13. P.C.H. Jordan and H.C. Longuet-Higgins, *Molec. Phys.* **5**, 121 (1962).
Semi-empirical calculation of low-lying levels.

14. F.O. Ellison, *J. phys. Chem., Ithaca,* **66**, 2294 (1962).
Semi-empirical calculation.

15. S. Fraga and B.J. Ransil, *J. chem. Phys.* **36**, 1127 (1962).
C.I.

16. E. Clementi and H. Clementi, *J. chem. Phys.* **36**, 2824 (1962).
Electron distributions.

17. D.M. Bishop and J.R. Hoyland, *Molec. Phys.* **7**, 161 (1963).
Single centre calculation

18. F.O. Ellison, *J. chem. Phys.* **43**, 3654 (1965).
Modified atoms in molecule.

19. R.M. Stevens and W.N. Lipscomb, *J. chemPhys.* **42**, 3666 (1965).
Magnetic properties.

20. A.J.A. Wu and F.O. Ellison, *Theor. Chim. Acta* **4**, 460 (1966).
Scaled united atom model.

21. F. Grimaldi, A. Lecourt, H. Lefebvre-Brion, and C.M. Moser, *J. molec. Spectrosc.* **20**, 341 (1966).
Rydberg levels.

22. P.E. Cade and W.M. Huo, *J. chemPhys.* **45**, 1063 (1966).
Dipole moment.

23. V. Magnasco and A. Perico, *J. chem. Phys.* **47**, 971 (1967).
Localized orbitals.

24. R.F.W. Bader, I. Keaveny, and P.E. Cade, *J. chem. Phys.* **47**, 3381 (1967).
Charge densities.

25. P.E. Cade and W.M. Huo, *J. chem. Phys.* **47**, 614 (1967).
Hartree–Fock limit.

26. J.R. de la Vega, Y. Fang, and H.F. Hameka, *Physica, 's Grav.* **36**, 577 (1967).
Diamagnetic susceptibility.

27. R.A. Hegstrom and W.N. Lipscomb, *J. chem. Phys.* **45**, 2378 (1966).
Magnetic properties.

28. R.A. Hegstrom and W.N. Lipscomb, *J. chem. Phys.* **46**, 1594 (1967).
Magnetic properties.

29. J.R. de la Vega and H.F. Hameka, *J. chem. Phys.* **47**, 1834 (1967).
Magnetic properties.

30. S.R. La Paglia, *Theor. Chim. Acta* **8**, 185 (1967).
Transition probabilities.

31. R.A. Hegstrom and W.N. Lipscomb, *J. chem. Phys.* **48**, 809, (1968).
Magnetic properties.

32. A.A. Frost, *J. phys. Chem., Ithaca,* **72**, 1289 (1968).
Floating spherical gaussian computation.

33. F. Grimaldi, A. Lecourt, and C.M. Moser, *Symp. Faraday Soc.* **2**, (1968).
Dipole moment from CI wave function.

34. C.F. Bender and E.R. Davidson, *Phys. Rev.* **183**, 23 (1969). Constrained natural orbitals + CI.

35. J.L. Harrison and L.C. Allen, *J. molec. Spectrosc.* **29**, 432 (1969). Also includes excited states $^1\Pi$, $^3\Pi$, $^1\Sigma^-$, $^3\Sigma^-$, $^1\Delta$, $^3\Sigma^+$, and $^3\Delta$.

BH^+

State	Internuclear distance	Basis set	Energy	Wave function	Reference
$X\,^2\Sigma^+$	2·296	Ext. STO	−24·82064	✓	1

Properties	*References*
Spin–orbit coupling constant of $A\,^2\Pi$	2

REFERENCES

1. P.E. Cade and W.M. Huo, *J. chem. Phys.* **47**, 614 (1967). Hartree–Fock limit.

2. T.E.H. Walker and W.G. Richards, *Symp. Faraday Soc.* **2**, 64 (1968). Spin–orbit coupling constant of $A\,^2\Pi$.

BN

State	Internuclear distance	Basis set	Energy	Wave function	Reference
$^1\Sigma^+$	2·421	Min. STO	−78·67870	✓	1
	2·354	Ext. STO	−78·902		3
$X\,^3\Pi$	2·421	Min. STO	−78·76138	✓	2
	2·465	Ext. STO			3
$^3\Sigma^+$	2·305	Ext. STO			3

Properties	*References*
Orbital energies	1, 2
Excited states	1, 2, 3
Potential curves	3
Spectroscopic constants	3
Dipole moment	1, 2

REFERENCES

1. J.L. Masse and M. Masse-Barlocher, *Helv. chim. Acta* **47**, 314 (1964). Min. STO.

2. J.L. Masse and M. Masse-Barlocher, *Helv. chim. Acta* **50**, 2560 (1967).
Min. STO

3. G. Verhaegen, W.G. Richards, and C.M. Moser, *J. chem. Phys.* **46**, 160 (1967).
Variational calculations on lowest states, predicts $^3\Pi$ ground state.

4. V.S. Nashpov and M.B. Khusidman, *Izv. Akad. Nauk S.S.S.R. Neorg. Material* **5**, 600 (1969).

Be_2

State	Internuclear distance	Basis set	Energy	Wave function	Reference
$X\,^1\Sigma_g^+$	3·78	Min. STO	−29·05825	✓	3
	3·78	Min. STO	−29·08671	✓	5
	3·78	Min. STO	−29·10537	✓	6

Properties	*References*
Repulsive potential curve	1, 5, 10
Population analysis	4, 6
Magnetic properties	7, 9
Charge density map	8
Orbital forces	8

REFERENCES

1. W.H. Furry and J.H. Bartlett, Jr., *Phys. Rev.* **38**, 1615 (1931).
Repulsive potential curve for $X\,^1\Sigma_g^+$ predicted from V.B. treatment.

2. W.H. Furry and J.H. Bartlett, Jr., *Phys. Rev.* **39**, 210 (1932).
Potential curves for excited states from V.B. treatment.

3. B.J. Ransil, *Rev. mod. Phys.* **32**, 239, 245 (1960).

4. B.J. Ransil and S. Fraga, *J. chem. Phys.* **34**, 727 (1961).
Minimum basis set–single determinant STO.

5. S. Fraga and B.J. Ransil, *J. chem. Phys.* **35**, 669 (1961).
Potential curves of ground and excited states discussed. $X\,^2\Sigma_u^+$ of Be_2^+ included. Best limited STO basis set used.

6. S. Fraga and B.J. Ransil, *J. chem. Phys.* **36**, 1127 (1962).
Limited CI included.

7. J.R. de la Vega and H.F. Hameka, *Physica, 's Grav.* **35**, 313 (1967).
B.J. Ransil wave functions (ref. 3) used.

8. R.F.W. Bader, W.H. Henneker, and P.E. Cade, *J. chem. Phys.* **46**, 3341 (1967).

9. J.R. de la Vega and H.F. Hameka, *J. chem. Phys.* **47**, 1834 (1967). B.J. Ransil wave functions (ref. 3) used to calculate rotational magnetic moments and anisotropies in the diamagnetic susceptibility.

10. C.F. Bender and E.R. Davidson, *J. chem. Phys.* **47**, 4972 (1967). Several excited states discussed.

11. V. Magnasco and A. Perico, *J. chem. Phys.* **47**, 971 (1967). Localization of molecular orbitals discussed. B.J. Ransil wave functions (ref. 3) used.

12. M.A. Marchetti and S.R. La Paglia, *J. chem. Phys.* **48**, 434 (1968). $^1\Sigma_g^+ - {}^1\Sigma_u^+$ dipole strengths from STO basis + CI. Wave functions as in ref. 3.

BeF

State	Internuclear distance	Basis set	Energy	Wave function	Reference
$^2\Sigma^+$	2·618	Ext. STO			1
$^2\Pi_r$	2·714	Ext. STO			1
$^2\Pi_i$	3·030	Ext. STO			1

Properties	*References*
Spectroscopic constants	1
Term values	1

REFERENCES

1. T.E.H. Walker and W.G. Richards, *Proc. phys. Soc.* **92**, 285 (1967).

BeH

State	Internuclear distance	Basis set	Energy	Wave function	Reference
$^2\Sigma^+$	2·538	Ext. STO	−15·15312	√	8
	2·538	Elliptical + CI	−15·221	√	10
	2·538	Constrained natural orbitals + CI	−15·2324	√	13

Properties	*References*
Dipole moment	7, 10, 13

Properties	*References*
Potential curves	8, 10
Ionization potential	8,
Spectroscopic constants	8
Charge distribution	9
Spin–orbit coupling	11, 12
Magnetic properties	13

REFERENCES

1. C.I. Ireland, *Phys. Rev.* **43**, 329 (1933).
Simplified V.B. calculation.
2. S. Aburto, R. Daudel, R. Gallardo, R. Lefebvre, and R. Mũnoz, *C.r. hebd. Séanc. Acad. Sci., Paris,* **247**, 1859 (1958).
Simplified min. basis.
3. S. Aburto, R. Gallardo, R. Mũnzo, R. Daudel, and R. Lefebvre, *J. Chim. phys.* **56**, 563 (1959).
Simplified min. basis.
4. F.O. Ellison, *J. phys. Chem., Ithaca,* **66**, 2294 (1962).
Semi-empirical V.B. calculation.
5. D.M. Bishop and J.R. Hoyland, *Molec. Phys.* **7**, 161 (1963).
One-centre calculation.
6. F.O. Ellison, *J. chem. Phys.* **43**, 3654 (1965).
Semi-empirical calculation.
7. P.E. Cade and W.M. Huo, *J. chem. Phys.* **45**, 1063 (1966).
Dipole moment.
8. P.E. Cade and W.M. Huo, *J. chem. Phys.* **47**, 614 (1967).
Hartree–Fock limit.
9. R.F.W. Bader, I. Keaveny, and P.E. Cade, *J. chem. Phys.* **47**, 3381 (1967).
Charge distribution.
10. A.C.H. Chan and E.R. Davidson, *J. chem. Phys.* **49**, 727 (1968).
Includes less accurate calculations of excited states.
11. T.E.H. Walker and W.G. Richards, *Symp. Faraday Soc.* **2**, 64 (1968).
12. T.E.H. Walker and W.G. Richards, *Phys. Rev.* **177**, 100 (1969.
Spin–orbit coupling constants.
13. C.F. Bender and E.R. Davidson, *Phys. Rev.* **183**, 23 (1969).

BeH^+

State	Internuclear distance	Basis set	Energy	Wave function	Reference
$^1\Sigma^+$	2·479	Min. STO	−14·8219	✓	1

State	Internuclear distance	Basis set	Energy	Wave function	Reference
	2·46	Ext. GTO	−14·829		6
	2·479	Ext. STO	−14·85396		4
	2·48	Elliptical + CI	−14·92196	✓	7

Properties	*References*
Potential curves	3
Electron affinity	6
Vibration frequency	6
Dipole moment	7

REFERENCES

1. I. Fischer, *Ark. Fys.* **5**, 349 (1952).
   Discussion of various approximations.
2. J.R. Miller, R.H. Friedman, R.P. Hirst, and F.A. Matsen, *J. chem. Phys.* **27**, 1385 (1959).
   Valence bond.
3. F. Jenč, *Colln Czech. chem. Commun. Engl. Edn.* **28**, 2064 (1963).
   Min. basis calculation.
4. P.E. Cade and W.M. Huo, *J. chem. Phys.* **47**, 614 (1967).
   Hartree–Fock limit.
5. W.A. Saunders and M. Krauss, *J. Res. natn. Bur. Stand.* **72A**, 85 (1968).
6. H. Preuss and R. Janoschek, *J. molec. Struct.* **3**, 423 (1969).
7. R.E. Brown, *J. chem. Phys.* **51**, 2879 (1969).

BeO

State	Internuclear distance	Basis set	Energy	Wave function	Reference
$X\,^1\Sigma^+$	2·515	Min. STO	−89·075	✓	1
	2·676	Double zeta	−89·4282		3
	2·5149	Ext. STO	−89·449	✓	4
	2·4377	Ext. STO	−89·451	✓	5, 6, 7
$a\,^3\Pi$	2·676	Double zeta	−89·4525		3
	2·7647	Ext. STO	−89·4877	✓	4

State	Internuclear distance	Basis set	Energy	Wave function	Reference
$A\,^1\Pi$	2·676	Double zeta	−89·4496		3
	2·7647	Ext. STO	−89·4835	✓	3, 4
$b\,^3\Sigma^+$	2·676	Double zeta	−89·40018		3
$B\,^1\Sigma^+$	2·676	Double zeta	−89·38143		3

Properties	*References*
Dipole moment	1, 2, 5, 6, 7
Quadrupole moment	2, 5, 6, 7
Spectroscopic constants	1, 2, 3, 4, 5, 6, 7
Excited-state constants	3, 4
Potential curves	2, 3, 4

REFERENCES

1. M. Bärlocher, *Helv. chim. Acta* **46**, 2920 (1963).
   Min. basis set calculation.
2. M. Yoshimine, *J. chem. Phys.* **40**, 2970 (1964).
   Same author later published better wave functions.
3. G. Verhaegen and W.G. Richards, *J. chem. Phys.* **45**, 1828 (1966).
   Includes discussion of nature of ground state.
4. W.M. Huo, K.F. Freed, and W. Klemperer, *J. chem. phys.* **46**, 3556 (1967).
   Discussion of nature of ground state and perturbations.
5. M. Yoshimine and A.D. McLean, *Int. J. Quantum Chem.* **1S**, 313 (1967).
6. M. Yoshimine and A.D. McLean, Supp. to *I.B.M. Jl Res. Dev.* (1964).
7. M. Yoshimine, *J. phys. Soc. Japan*, **25**, 110 (1968).
   Refs. 5, 6, and 7 all refer to the same wave function, which is close to Hartree–Fock limit.

BeS

State	Internuclear distance	Basis set	Energy	Wave function	Reference
$X\,^1\Sigma^+$	3·3	Ext. STO	−412·0972		1
$^3\Pi$	3·3	Ext. STO	−412·0835		1
$A\,^1\Pi$	3·3	Ext. STO	−412·0729		1
$^3\Sigma^+$	3·3	Ext. STO	−412·0287		1

State	Internuclear distance	Basis set	Energy	Wave function	Reference
$B\,^1\Sigma^+$	3·3	Ext. STO	(–411·978)		1

Properties	*References*
Term values	1

REFERENCE

1. G. Verhaegen and W.G. Richards, *Proc. phys. Soc.* **90**, 579 (1967).

$\underline{C_2}$

State	Internuclear distance	Basis set	Energy	Wave function	Reference
$X\,^1\Sigma_g^+$	2·3481	Guassian lobe	–75·40620		9
	2·3475	Min. STO +CI	–75·31929		3
	2·3475	Min. STO	–75·22381	√	1
	2·3475	STO	–75·2545	√	15

Properties	*References*
Orbital energies	1, 4, 15
Population analysis	2, 3, 4
Charge density contours	5, 10
Potential curves	6, 8, 13
Spectroscopic constants	6, 8, 13
Term values	6
Magnetic properties	7, 11
Dissociation energy	10
Orbital forces	10

REFERENCES

1. B.J. Ransil, *Rev. mod. Phys.* **32**, 239, 245 (1960).

2. S. Fraga and B.J. Ransil, *J. chem. Phys.* **34**, 727 (1961). Wave functions as in ref. 1.

3. S. Fraga and B.J. Ransil, *J. chem. Phys.* **36**, 1127 (1962). Wave functions as in ref. 1. Limited CI included.

4. E. Clementi and H. Clementi, *J. chem. Phys.* **36**, 2824 (1962). Wave functions as in ref. 1.

5. A.C. Wahl, *Science, N.Y.*, **151**, 961 (1966).
STO basis set.

6. P.F. Fougere and R.K. Nesbet, *J. chem. Phys.* **44**, 285 (1966).
Many states investigated using different STO basis sets, including CI.

7. J.R. de la Vega and H.F. Hameka, *Physica, 's Grav.* **35**, 313 (1967).
Wave functions as in ref. 1.

8. G. Verhaegen, W.G. Richards, and C.M. Moser, *J. chem. Phys.* **46**, 160 (1967).
Ext. STO basis set used. $^3\Pi_u$ predicted as ground state.

9. R.J. Buenker, S.D. Peyerimhoff, and J.L. Whitten, *J. chem. Phys.* **46**, 2029 (1967).

10. R.F.W. Bader, W.H. Henneker, and P.E. Cade, *J. chem. Phys.* **46**, 3341 (1967).
J.B. Greenshields wave functions (unpub.) used.

11. J.R. de la Vega and H.F. Hameka, *J. chem. Phys.* **47**, 1834 (1967).
Wave functions as in ref. 1.

12. M.A. Marchetti and S.R. La Paglia, *J. chem. Phys.* **48**, 434 (1968).
$^1\Sigma_g^+ - ^1\Sigma_u^+$ dipole strengths from wave functions in ref. 1.

13. G. Verhaegen, *J. chem. Phys.* **49**, 4696 (1968).
Ext. STO basis + CI predicts $^3\Pi_u$ ground state.

14. S.R. La Paglia, *J. molec. Spectrosc.* **24**, 302 (1967).
Oscillator strengths for $^1\Sigma_g^+ - ^1\Sigma_u^+$ using CI. Wave functions as in ref. 1.

15. E. Clementi, *Gazz. chim. ital.* **91**, 717 (1961).

C_2^+

State	Internuclear distance	Basis set	Energy	Wave function	Reference
$^4\Sigma_g^-$	2·6	Ext. STO + CI	−75·138	✓	1

Properties	*References*
Term values	1
Orbital energies	1
Spectroscopic constants	1

REFERENCE

1. G. Verhaegen, *J. chem. Phys.* **49**, 4696 (1968).
$^4\Sigma_g^-$, $^2\Pi_u$, $^2\Sigma_u^+$, $^4\Pi_g$ states included.

CH

State	Internuclear distance	Basis set	Energy	Wave function	Reference
$^2\Pi_r$		Min. STO	−38·164	✓	2
	2·086	Ext. STO	−38·27958	✓	15
	2·124	Constrained orbitals + CI	−38·4399		21
$^2\Sigma^-$			−38·120		10

Properties	*References*
Excitation energies	2, 3
Dipole moment	2, 3, 14, 16
Dissociation energy	2, 3, 15
Nuclear spin–spin coupling	13
Spectroscopic constants	15
Potential curves	15
Ionization potential	15
Charge distribution	16, 21
Electron affinity	18
Spin–orbit coupling constants	19, 20
Magnetic properties	21

REFERENCES

1. J. Higuchi, *J. chem. Phys.* **22**, 1339 (1954).
   Simplified calculations on several states.

2. M. Krauss, *J. chem. Phys.* **28**, 1021 (1958).
   Min. basis set.

3. M. Krauss and J.F. Wehner, *J. chem. Phys.* **29**, 1287 (1958).
   CI on wave function of ref. 2.

4. A.C. Hurley, *Proc. R. Soc.* **A248** , 119 (1958).
   CI on wave function of ref. 2.

5. A.C. Hurley, *Proc. R. Soc.* **A249**, 402 (1959).
   Excited states by using semi-empirical correction.

6. M. Roux and J.L. Masse, *J. Chim. phys.* **56**, 834 (1959).
   Includes min. basis calculation of excited state.

7. J.L. Masse, *J. Chim. phys.* **58**, 572 (1961).
   Includes min. basis calculation of excited state.

8. M. Cornille, *Cah. Phys.* **14**, 497 (1960).
   Charge density.

9. H. Ben Jemia and R. Lefebvre, *J. Chim. phys.* **58**, 306 (1961).
   Charge density.

10. J.M. Foster and S.F. Boys, *Rev. mod. Phys.* **32**, 305 (1960).

11. P.C.H. Jordan and H.C. Longuet-Higgins, *Molec. Phys.* **5**, 121 (1962).
Semi-empirical.

12. D.M. Bishop and J.R. Hoyland, *Molec. Phys.* **7**, 161 (1963).
Single configuration V.B.

13. D.S. Barstow and J.W. Richardson, *J. chem. Phys.* **42**, 4018 (1965).
Nuclear spin–spin coupling from min. basis calculation.

14. P.E. Cade and W.M. Huo, *J. chem. Phys.* **45**, 1063 (1966).
Dipole moment.

15. P.E. Cade and W.M. Huo, *J. chem. Phys.* **47**, 614 (1967).
Hartree–Fock limit.

16. R.F.W. Bader, I. Keaveny, and P.E. Cade, *J. chem. Phys.* **47**, 3381 (1967).
Charge distribution

17. N. Grün, *Z. Naturf.* **22a**, 1928 (1967).
Valence bond.

18. P.E. Cade, *Proc. phys. Soc.* **91**, 842 (1967).
Electron affinity.

19. T.E.H. Walker and W.G. Richards, *Symp. Faraday Soc.* **2**, 64 (1968).
Spin–orbit coupling.

20. T.E.H. Walker and W.G. Richards, *Phys. Rev.* **177**, 100 (1969).
Spin–orbit coupling.

21. C.F. Bender and E.R. Davidson, *Phys. Rev.* **183**, 23 (1969).

CH^+

State	Internuclear distance	Basis set	Energy	Wave function	Reference
$^1\Sigma^+$		Min. STO	−37·779	✓	1
	2·137	Ext. STO	−37·90881	✓	4
	2·2356	Valence bond	−37·91786	✓	3

Properties	*References*
Excitation energies	1, 3
Dipole moment	1
Atomic populations	1

REFERENCES

1. M. Krauss, *J. chem. Phys.* **28**, 1201 (1958).
Min. basis set calculation.

2. F. Jenč, *Colln. Czech. chem. Commun. Engl. Edn.* **28**, 2064 (1963).
Min. basis set calculation.

3. P.L. Moore, J.C. Browne, and F.A. Matson, *J. chem. Phys.* **43**, 903 (1965).
Generalized V.B. calculations using many 'structures'.

4. P.E. Cade and W.M. Huo, *J. chem. Phys.* **47**, 614 (1967).
Hartree–Fock limit.

CH^-

State	Internuclear distance	Basis set	Energy	Wave function	Reference
$^3\Sigma^-$	2·086	Ext. STO	–38·29003		1

Property	*Reference*
Electron affinity	1

REFERENCE

1. P.E. Cade, *Proc. phys. Soc.* **91**, 842 (1967).
Hartree–Fock limit.

CN^-

State	Internuclear distance	Basis set	Energy	Wave function	Reference
$^1\Sigma^+$	2·18	STO	–91·9273		1
	2·935	GTO	–92·1016		2
	2·1791	Double zeta	–92·4123		4

Properties	*References*
Orbital energies	1, 4
Charge density	1
Force	1
Dipole moment	1, 4
Quadrupole moment	1, 3, 4
Proton affinity	2, 4

REFERENCES

1. R. Bonaccorsi, C. Petrongolo, E. Scrocco, and J. Tomasi, *J. chem. Phys.* **48**, 1500 (1968).

2. A.C. Hopkinson, N.K. Holbrook, K. Yates, and I.G. Csizmadia, *J. chem. Phys.* **49**, 3596 (1968).
Proton affinities with a variety of basis sets.

3. R. Bonaccorsi, E. Scrocco, and J. Tomasi, *J. chem. Phys.* **50**, 2940 (1969).
Electric field gradient at ^{14}N.

4. R. Bonaccorsi, C. Petrongolo, E. Scrocco, and J. Tomasi, *Chem. Phys. Lett.* **3**. 473 (1969).

5. T.A. Claxton, *Chem. Phys. Lett.* **4**, 469 (1969).
Hyperfine constants of CN radical.

CO

State	Internuclear distance	Basis set	Energy	Wave function	Reference
$^1\Sigma^+$	2·132	Min. STO	−112·32604		6
	2·132	Min. STO	−112·32705	✓	3
	2·132	Min. STO	−112·3276	✓	11
	2·132	Min. STO	−112·392		22
	2·132	Min. STO	−112·39602	✓	12
	2·132	SCGF	−112·4019		33
	2·132	Min. STO +CI	−112·4385	✓	7
	2·132	GTO	−112·4680		32
	2·1319	Min. STO	−112·5752	✓	10
	2·132	Ext. STO	−112·7588	✓	17
	2·132	Contr. GTO	−112·7622		36
	2·132	Ext. STO	−112·7878		35
	2·081	Ext. STO	−112·7879	✓	19
	2·132	Ext. STO	−112·7891	✓	30
a $^3\Pi$	2·2853	Ext. STO	−112·5742	✓	21

Properties	*References*
Orbital energies	1, 3, 6, 19, 21, 22, 30
Charge density	1, 19
Dipole moment	1, 2, 3, 6, 7, 12, 15, 17, 19, 20, 21, 25, 29, 34, 35, 36, 37
Quadrupole coupling	2, 8, 15, 17, 21, 25
Atomic population	2, 13, 18, 24
Excited states	3, 4, 5, 10, 11, 18, 34

Properties	*References*
Potential curves	4, 10, 17, 19, 20, 22, 30
Magnetic properties	9, 14, 27, 28, 35, 36
Polarizability	16, 23, 25, 36
Spectroscopic constants	17, 19, 20, 22
Localized orbitals	26
Transition probabilities	31
Proton affinity	32
Scattering cross-section	38

REFERENCES

1. R.C. Sahni, *Trans. Faraday Soc.* **49**, 1246 (1953). Min. STO.

2. B.J. Ransil, *J. chem. Phys.* **30**, 1113 (1959). Dipole and quadrupole coupling constants.

3. H. Brion and C.M. Moser, *J. chem. Phys.* **32**, 1194 (1960). Min. basis set used for excited states.

4. P. Merryman, C.M. Moser, and R.K. Nesbet, *J. chem. Phys.* **33**, 631 (1960). Min. basis set for excited states.

5. H. Lefebvre-Brion, C.M. Moser, and R.K. Nesbet, *J. chem. Phys.* **33** (1960). Min. basis set for excited $^1\Sigma^+$ states.

6. B.J. Ransil, *Rev. mod. Phys.* **32**, 239 (1960).

7. A.C. Hurley, *Rev. mod. Phys.* **32**, 400 (1960). 48-term CI.

8. J.W. Richardson, *Rev. mod. Phys.* **32**, 461 (1960). Quadrupole coupling constant.

9. B.J. Ransil and S. Fraga, *J. chem. Phys.* **34**, 727 (1961). Magnetic shielding.

10. H. Lefebvre-Brion, C.M. Moser, and R.K. Nesbet, *J. chem. Phys.* **34**, 1950 (1961). Excited states.

11. H. Lefebvre-Brion, C.M. Moser, and R.K. Nesbet, *J. chem. Phys.* **35**, 1702 (1961). Excited states.

12. S. Fraga and B.J. Ransil, *J. chem. Phys.* **36**, 1127 (1962). Limited CI also carried out.

13. E. Clementi and H. Clementi, *J. chem. Phys.* **36**, 2824 (1962). Atomic population.

14. M. Karplus and H.J. Kolker, *J. chem. Phys.* **38**, 1263 (1963). Magnetic susceptibility by perturbation theory.

15. H. Brion, C.M. Moser, R.K. Nesbet, and M. Yamazaki, *J. chem. Phys.* **38**, 2311 (1963).
Electric field gradient and dipole moment.

16. M. Karplus and H.J. Kolker, *J. chem. Phys.* **39**, 2011 (1963).
Electric polarizability.

17. R.K. Nesbet, *J. chem. Phys.* **40**, 3619 (1964).
Ext. STO.

18. H. Lefebvre-Brion, C.M. Moser, and R.K. Nesbet, *J. molec. Spectrosc.* **13**, 418 (1964).
Excited states.

19. W.M. Huo, *J. chem. Phys.* **43**, 624 (1965).
Near Hartree–Fock limit.

20. R.K. Nesbet, *J. chem. Phys.* **43**, 4403 (1965).
Excited states.

21. W.M. Huo, *J. chem. Phys.* **45**, 1554 (1966).
Probably near Hartree–Fock limit, results also for $^1\Pi$.

22. R.C. Sahni, C.D. La Budda, and B.C. Sawhney, *Trans. Faraday Soc.* **62**, 1993 (1966).
Min. STO, RKR method used for potential curves.

23. J.M. O'Hare and R.P. Hurst, *J. chem. Phys.* **46**, 2356 (1967).
Polarizability.

24. E.R. Davidson, *J. chem. Phys.* **46**, 3320 (1967).
Population analysis.

25. A.D. McLean and M. Yoshimine, *J. chem. Phys.* **46**, 3682 (1967).

26. V. Magnasco and A. Perico, *J. chem. Phys.* **47**, 971 (1967).
Localized orbitals.

27. J.R. de la Vega and H.F. Hameka, *J. chem. Phys.* **47**, 1834 (1967).
Magnetic properties.

28. J.R. de la Vega, D. Ziobro, and H.F. Hameka, *Physica, 's Grav.* **37**, 265 (1967).
Magnetic susceptibility.

29. F. Grimaldi, A. Lecourt, and C. Moser, *Int. J. Quantum Chem.* **1S**, 153 (1967).
Dipole moment from CI – gives correct sign.

30. A.D. McLean and M. Yoshimine, Supp. to *I.B.M. Jl Res. Dev.* (1967).
Very near Hartree–Fock limit.

31. S.R. La Paglia, *Theor. Chim. Acta* **8**, 185 (1967).
Transition probabilities.

32. A.C. Hopkinson, N.K. Holbrook, K. Yates, and I.G. Csizmadia, *J. chem. Phys.* **49**, 3596 (1968).
Gaussian basis proton affinity calculations.

33. M. Klessinger, *Symp. Faraday Soc.* **2**, 73 (1968).
Includes some CI.

34. A Julg and A. Pellegatti, *J. Chim. phys.* **65**, 242 (1968).
Dipole moment of excited states.

35. R.M. Stevens and M. Karplus, *J. chem. Phys.* **49**, 1094 (1968).
Variation of dipole with internuclear distance.

36. D.B. Neuman and J.W. Moskowitz. *J. chem. Phys.* **50**, 2216 (1969).
Contracted Gaussians. One electron properties given.

37. I.H. Hillier and V.R. Saunders, *Chem. Phys. Lett.* **4**, 163 (1968).
Gaussian basis.

38. D.A. Kohl, *J. chem. Phys.* **51**, 2896 (1969).

CO^+

State	Internuclear distance	Basis set	Energy	Wave function	Reference
$X\ ^2\Sigma^+$	2·132	Min. STO	–111·9543	✓	1
$A\ ^2\Pi$	2·132	Min. STO	–111·8938	✓	1
$B\ ^2\Sigma^+$	2·132	Min. STO	–111·5815	✓	1

Properties	*References*
Spectroscopic constants	1
Potential curves	1
Orbital energies	1
Spin–orbit coupling	2

REFERENCES

1. R.C. Sahni, B.C. Sawhney, *Trans. Faraday Soc.* **63**, 1 (1967).
Min. STO calculations.

2. T.E.H. Walker and W.G. Richards, *Symp. Faraday Soc.* **2**, 64 (1968).
Calculation of spin–orbit coupling constant for $A\,^2\Pi$ state.

CS

State	Internuclear distance	Basis set	Energy	Wave function	Reference
$X\ ^1\Sigma^+$	2·89964	Ext. STO	–435·3297	✓	1

Properties	*References*
Orbital energies	1, 2
Dipole moment	1

REFERENCES

1. W.G. Richards, *Trans. Faraday Soc.* **63**, 257 (1967). Discussion of structure of thiocarbonyl complexes.

2. I.H. Hillier and V.R. Saunders, *Chem. Phys. Lett.* **4**, 163 (1969). Gaussian basis.

CaO

State	Internuclear distance	Basis set	Energy	Wave function	Reference
$^1\Sigma^+$	3·3412	STO	−751·5478	✓	1, 2, 3
	3·4412	STO	−751·5591		2

Properties	*References*
Orbital energies	1, 3
Potential curve	1
Dipole moment	2, 3
Dissociation energy	2
Magnetic properties	3
Spectroscopic constants	3

REFERENCES

1. A.D. McLean and M. Yoshimine, Supp. to *I.B.M. Jl Res. Dev.* (1967).

2. M. Yoshimine and A.D. McLean, *Int. J. Quantum Chem.* **1S**, 313 (1967).

3. M. Yoshimine, *J. phys. Soc. Japan*, **25**, 1100 (1968). Calculations approach Hartree–Fock limit.

ClH

State	Internuclear distance	Basis set	Energy	Wave function	Reference
$^1\Sigma^+$	2·404	One-centre STO	−458·8378		1
	2·404	One-centre STO	−458·8378		2
		Min. STO	−459·421		4
	2·4085	STO	−459·8037	✓	3
	2·4087	Ext. STO	−460·1103	✓	8
	2·4087	Ext. STO	−460·1119	✓	7
	2·4087	Ext. STO	−460·1119	✓	9

Properties	*References*
Dipole moment	1, 2, 3, 4, 5, 9, 11
Potential curve	3
Spectroscopic constants	3, 8
Excited states	3
Population analysis	4
Quadrupole coupling	4, 6, 9, 10
Orbital energies	8, 9
Ionization potential	8
Magnetic properties	9
Force constant	12
Charge distribution	13

REFERENCES

1. R. Moccia, *J. chem. Phys.* **37**, 910 (1962).
One-centre calculation.

2. R. Moccia, *J. chem. Phys.* **40**, 2186 (1964).
One-centre calculation.

3. R.K. Nesbet, *J. chem. Phys.* **41**, 100 (1964).
Excited states studied by virtual orbitals.

4. E. Scrocco and J. Tomasi, *Molecular Orbitals in Chemistry, Physics, and Biology* (ed. Löwdin and Pullman), p. 263. Academic Press, New York, 1964.
Uses factorization technique.

5. P.E. Cade and W.M. Huo, *J. chem. Phys.* **45**, 1063 (1966).
Dipole moment.

6. P. Pyykko, *Proc. phys. Soc.* **92**, 841 (1967).
Deuteron quadrupole coupling constant.

7. A.D. McLean and M. Yoshimine, Supp. to *I.B.M. Jl Res. Dev* (1967).
Very near Hartree–Fock Limit.

8. P.E. Cade and W.M. Huo, *J. chem. Phys.* **47**, 649 (1967).
Near Hartree–Fock Limit.

9. A.D. McLean and M. Yoshimine, *J. chem. Phys.* **47**, 3256 (1967).
Very near Hartree–Fock Limit.

10. R.M. Dixon and J.A. Smith, *Trans. Faraday Soc.* **64**, 1 (1968).
Deuteron quadrupole coupling constant.

11. F. Grimaldi, A. Lecourt, and C.M. Moser, *Symp. Faraday Soc.* **2**, 59 (1968).
Dipole moment with CI.

12. R.F.W. Bader and J.L. Ginsberg, *Can. J. Chem.* **47**, 3061 (1969).
Force constant.

13. P.E. Cade, R.F.W. Bader, W.H. Henneker, and I. Keaveny, *J. chem. Phys.* **50**, 5313 (1969).
Charge distribution.

14. H. Preuss and R. Janoschek, *J. molec. Struct.* **3**, 423 (1969).
Preliminary calculation, $E = -453{\cdot}0534$.

ClH^+

State	Internuclear distance	Basis set	Energy	Wave function	Reference
$^2\Sigma^+$	2·485	Ext. STO	−459·6762		1

REFERENCE

1. P.E. Cade and W.M. Huo, J. *chem. Phys.* **47**, 649 (1967).
Near Hartree–Fock limit.

F_2

State	Internuclear distance	Basis set	Energy	Wave function	Reference
$X\,^1\Sigma_g^+$	2·680	Min. STO + CI	−197·87017	✓	1
	2·680	Min. STO	−197·87694	✓	2
	2·517	Min. STO	−197·88749		6
	2·680	Min. STO + CI	−197·9558	✓	7
	2·661	GTO	−198·1137		25
	2·684	V.B. GTO	−198·13004		24
	2·68	Perturbed Hartree–Fock: Ext. STO	−198·73107	✓	12
	2·68	Min. STO	−198·76825		18
	2·68	Ext. STO	−198·7683	✓	11
	3·1	STO + CI	−198·81788		22
	3·0	Ext. Hartree–Fock Valence: Ext. STO	−198·8378	✓	13
$^3\Pi_u$	2·918	Min. STO	−197·73612		6
$^1\Pi_u$	2·948	Min. STO	−197·67288		6

State	Internuclear distance	Basis set	Energy	Wave function	Reference
$^3\Pi_g$	3·266	Min. STO	–197·66548		6
$^1\Pi_g$	3·281	Min. STO	–197·60925		6
$^3\Sigma_g^-$	3·589	Min. STO	–197·59391		6
$^1\Delta_g$	3·599	Min. STO	–197·53887		6
$^1\Sigma_g^+$	3·610	Min. STO	–197·48384		6

Properties	*References*
O.bital energies	1, 2, 11, 15
Ionization potentials	2, 25
Term values	3, 4, 6
Dissociation energy	3, 4, 17, 24
Potential curve	4, 6, 13, 22, 24
Population analysis	5
Spectroscopic constants	6, 11, 13, 25
Magnetic properties	8, 9, 10, 12, 20, 21
Quadrupole moment	11
Charge densities	11, 12, 14, 17, 18
Electric polarizability	12, 16
Spin rotation	12
Orbital forces	17

REFERENCES

1. J. Eve, *Proc. R. Soc.* **A246**, 582 (1958).
2. B.J. Ransil, *Rev. mod. Phys.* **32**, 239, 245 (1960).
3. K. Hijikata, *Rev. mod. Phys.* **32**, 445 (1960).
   Pilot calculation for ref. 4.
4. K. Hijikata, *J. chem. Phys.* **34**, 221, 231 (1961).
   Min. STO on states $X\ ^1\Sigma_g^+$, $^3\Pi_u$, $^1\Pi_u$, $^3\Pi_g$, $^1\Pi_g$, $^3\Sigma_u^+$, $^3\Sigma_u^-$, $^1\Sigma_g^+$, $^1\Sigma_u^-$, $^1\Delta_g$, $^3\Delta_u$.
   Wave functions given.
5. S. Fraga and B.J. Ransil, *J. chem. Phys.* **34**, 727 (1961).
   Wave functions given in ref. 2.
6. S. Fraga and B.J. Ransil, *J. chem. Phys.* **35**, 669 (1961).
   Wave functions given in ref. 2.
7. S. Fraga and B.J. Ransil, *J. chem. Phys.* **36**, 1127 (1962).
   Limited CI

8. C.W. Kern and W.N. Lipscomb, *J. chem. Phys.* **37**, 260 (1962). Wave functions given in ref. 2.

9. M. Karplus and H.J. Kolker, *J. chem. Phys.* **38**, 1263 (1963). Variation, perturbation calculation using wave functions in ref. 2.

10. H. J. Kolker and M. Karplus, *J. chem. Phys.* **41**, 1259 (1964). Variation, perturbation calculation using wave functions in ref. 2.

11. A.C. Wahl, *J. chem. Phys.* **41**, 2600 (1964).

12. R.M. Stevens and W.N. Lipscomb, *J. chem. Phys.* **41**, 3710 (1964). Symmetry restrictions on '*u*' and '*g*'.

13. G. Das and A.C. Wahl, *J. chem. Phys.* **44**, 87 (1966).

14. A.C. Wahl, *Science, N.Y.* **151**, 961 (1966). STO basis.

15. R.J. Buenker, S.D. Peyerimhoff, and J.L. Whitten, *J. chem. Phys.* **46**, 2029 (1967). Correlation diagram with C_2, N_2, and O_2 included. Wave functions given in ref. 11.

16. J.M. O'Hare and R.P. Hurst, *J. chem. Phys.* **46**, 2356 (1967). Perturbation method using wave functions in ref. 2.

17. R.F.W. Bader, W.H. Henneker, and P.E. Cade, *J. chem. Phys.* **46**, 3341 (1967). Wave functions given in ref. 11.

18. B.J. Ransil and J.J. Sinai, *J. chem. Phys.* **46**, 4050 (1967). Wave functions given in refs. 2 and 11.

19. V. Magnasco and A. Perico, *J. chem. Phys.* **47**, 971 (1967). Localization of atomic and molecular orbitals using wave functions in ref. 2.

20. J.R. de la Vega and H.F. Hameka, *J. chem. Phys.* **47**, 1834 (1967). Wave functions given in ref. 2.

21. J.R. de la Vega and H.F. Hameka, *Physica, 's Grav.* **35**, 313 (1967) Wave functions given in ref. 2.

22. F.E. Harris and H.H. Michels, *Int. J. Quantum Chem* **1S**, 329 (1967). Double zeta basis set.

23. M.A. Marchetti and S.R. La Paglia, *J. chem. Phys.* **48**, 434 (1968). ${}^1\Sigma_u^+ - {}^1\Sigma_g^+$ dipole strengths from wave function in ref. 2 + CI.

24. C.G. Balint-Kurti and M. Karplus, *J. chem. Phys.* **50**, 478 (1969). Correlation energy estimated.

25. R. Janoschek and H. Preuss, *Z. Naturf.* **24A**, 674, (1969).

26. H. Preuss and R. Janoschek, *J. molec. Struct.* **3**, 423 (1969). Preliminary calculation, $E = -198{\cdot}1137$.

F_2^+

State	Internuclear distance	Basis set	Energy	Wave function	Reference
$X\,^2\Pi_g$	2·415	Min. STO	–197·43193		2
$^2\Sigma_g^+$	2·984	Min. STO	–197·34273		2
$^4\Sigma_u^-$	2·771	Min. STO	–197·30307		2
$^2\Delta_u$	2·784	Min. STO	–197·21394		2
$^2\Sigma_u^-$	2·805	Min. STO	–197·20556		2
$^2\Sigma_u^+$	2·792	Min. STO	–197·15699		2
$^2\Pi_g$	3·332	Min. STO	–197·02812		2

Properties	*References*
Potential curve	1, 2
Spectroscopic constants	2
Term values	1, 2

REFERENCES

1. K. Hijikata, *J. chem. Phys.* **34**, 231 (1961).
Calculations on $^2\Pi_g$, $^2\Pi_u$, $^2\Sigma_u^+$, $^2\Sigma_u^-$ states.

2. S. Fraga and B.J. Ransil, *J. chem Phys.* **35**, 669 (1961).

3. K. Hijikata, *Rev. mod. Phys.* **32**, 445 (1960).
Pilot calculation on $^2\Pi_g$ $^2\Pi_u$ $^2\Sigma_u^+$ $^2\Sigma_u^-$ states. See ref. 1.

F_2^{++}

State	Internuclear distance	Basis set	Energy	Wave function	Reference
$X\,^3\Sigma_g^-$	2·389	Min. STO	–196·36018		1

Properties	*References*
Spectroscopic constants	1
Term values	1

REFERENCES

1. S. Fraga and B.J. Ransil, *J. chem. Phys.* **35**, 669 (1961).

F_2^-

State	Internuclear distance	Basis set	Energy	Wave function	Reference
$^2\Sigma_u^+$	2·68345	V.B.	−198·43924		1

Properties	*Reference*
Potential curve	1

REFERENCE

1. C.G. Balint-Kurti and M. Karplus, *J. chem. Phys.* **50**, 478 (1969). Correlation energies estimated.

FH

State	Internuclear distance	Basis set	Energy	Wave function	Reference
$X\,^1\Sigma^+$	1·7328	Min. STO	−99·475	√	8
	1·7328	Min. STO	−99·479		11
	1·7328	Min. STO	−99·48631	√	7
	2·0	Group functions	−99·51889		62
	1·733	Min. STO	−99·56401	√	20
	1·7328	Valence bond	−99·6271	√	31
	1·7052	Valence bond	−99·6651	√	30
		Ext. STO	−99·96339		12
	2·0	Double zeta	−100·0031	√	59
	1·7328	One centre	−100·0053		34
	1·75	Ext. GTO	−100·0178		36
	1·7324	Ext. GTO	−100·0356		63
	1·7328	Ext. STO	−100·0571		23
	1·7328	Ext. STO	−100·0580	√	24
	1·7328	Ext. GTO	−100·0622		42
	1·696	Ext. STO	−100·0708	√	50
	1·7328	Ext. STO + CI	−100·3564		65

Properties	*References*
Dipole moment	1, 8, 11, 12, 13, 20, 23, 24, 25, 34, 40, 42, 57, 61, 64, 65

Properties	*References*
Spectroscopic constants	1,8,13,17,23,45,50,57,61, 62,65,66
Excitation energies	8,17,23,24
Charge density	16,32,37,39,45,51,52
Magnetic properties	18,21,22,26,27,28,33,35, 41,43,47,54,57,65
Deuteron coupling constant	19, 56, 60
Rotational *g* factor	47
Nuclear spin–spin coupling	48
Hyperpolarizabilities	49
Ionization potential	50
Transition probabilities	55,61
Proton affinity	63

REFERENCES

1. Z. Náray, *Acta phys. hung.* **3**, 255 (1953). Spectroscopic constants.
2. D. Kastler, *J. Chim. phys.* **50**, 556 (1953).
3. D. Kastler, *C.r. hebd. Séanc. Acad. Sci., Paris*, **236**, 1271 (1953).
4. D. Kastler, *C.r. hebd. Séanc. Acad. Sci., Paris*, **236**, 1359 (1953). Min. basis V.B. calculations and improvements with CI.
5. A.B.F. Duncan, *J. Am. chem. Soc.* **77**, 2107 (1955). Min. basis set.
6. H. Hamano, *Bull. chem. Soc. Japan* **30**, 741 (1957). Min. basis valence electron calculation.
7. A.M. Karo and L.C. Allen, *J. Am. chem. Soc.* **80**, 4496 (1958). Min. basis set + CI.
8. M. Krauss, *J. chem. Phys.* **28**, 1021 (1958). Min. basis set.
9. M. Krauss and J.F. Wehner, *J. chem. Phys.* **29**, 1287 (1958). CI on wave function of ref. 8.
10. R. Gáspár and I. Tamássy, *Acta. phys. hung.* **9**, 105 (1958). United atom.
11. R.A. Ballinger, *Molec. Phys.* **2**, 139 (1959). Min. basis set.
12. A.M. Karo and L.C. Allen, *J. chem. Phys.* **31**, 968 (1959). Includes CI.

13. B.J. Ransil, *Rev. mod. Phys.* **32**, 239 (1960).
Spectroscopic constants.

14. B.J. Ransil, *Rev. mod. Phys.* **32**, 245 (1960).
Min. basis set.

15. R.K. Nesbet, *Rev. mod. Phys.* **32**, 272 (1960).
Preliminary results.

16. S. Fraga and B.J. Ransil, *J. chem. Phys.* **34**, 727 (1961).
Population analysis.

17. S. Fraga and B.J. Ransil, *J. chem. Phys.* **35**, 669 (1961).
Spectroscopic constants.

18. C.W. Kern and W.N. Lipscomb, *J. chem. Phys.* **37**, 260 (1962).
Magnetic shielding.

19. H.J. Kolker and M. Karplus, *J. chem. Phys.* **36**, 960 (1962).
Deuteron coupling constant.

20. S. Fraga and B.J. Ransil, *J. chem. Phys.* **36**, 1127 (1962).
CI wave function.

21. R.P. Hurst, M. Karplus, and T.P. Das, *J. chem. Phys.* **36**, 2786 (1962).
Susceptibility.

22. T.P. Das and M. Karplus, *J. chem. Phys.* **36**, 2275 (1962).
Magnetic properties.

23. R.K. Nesbet, *J. chem. Phys.* **36**, 1518 (1962).
Near Hartree–Fock limit.

24. E. Clementi, *J. chem. Phys.* **36**, 33 (1962).
Close to Hartree–Fock limit.

25. R. Moccia, *J. chem. Phys.* **37**, 910 (1962).
One centre.

26. M. Karplus and H.J. Kolker, *J. chem. Phys.* **38**, 1263 (1963).
Magnetic susceptibility.

27. M. Karplus and H.J. Kolker, *J. chem. Phys.* **39**, 2011 (1963).
Electric polarizability.

28. H. Hamano, H. Kim, and H.F. Hameka, *Physica, s' Grav.* **29**, 111 (1963).
Magnetic properties.

29. A. Mukharji and M. Karplus, *J. chem. Phys.* **38**, 44 (1963).

30. D.M. Bishop and J.R. Hoyland, *Molec. Phys.* **7**, 161 (1963).
Single centre.

31. S. Peyerimhoff, *Z. Naturf.* **18a**, 1197 (1963).
Valence bond.

32. R.F.W. Bader and G.A. Jones, *Can. J. Chem.* **41**, 2251 (1963).
Electron density.

33. H.J. Kolker and M. Karplus, *J. chem. Phys.* **41**, 1259 (1964).
Nuclear magnetic shielding.

34. R. Moccia, *J. chem. Phys.* **40**, 2164 (1964).
One centre.

35. R.M. Stevens and W.N. Lipscomb, *J. chem. Phys.* **41**, 184 (1964).
Magnetic properties and polarizability.

36. M.C. Harrison, *J. chem. Phys.* **41**, 499 (1964).
Gaussian basis.

37. C.W. Kern and M. Karplus, *J. chem. Phys.* **40**, 1374 (1964). and **43**, 2926 (1965).
Analysis of wave functions.

38. K. Pecul, *Acta phys. pol.* **27**, 713 (1965).
Min. basis V.B.

39. R.F.W. Bader and W.H. Hanneker, *J. Am. chem. Soc.* **88**, 280 (1966).
Charge density.

40. P.E. Cade and W.M. Huo, *J. chem. Phys.* **45**, 1063 (1966).
Dipole moment.

41. D. Zeroka and H.F. Hameka, *J. chem. Phys.* **45**, 300 (1966).
Magnetic shielding constants.

42. J. Moskowitz, D. Neumann and M.C. Harrison, in *Quantum theory of atoms, molecules and the solid state* (ed Löwdin), p. 227. Academic Press, New York, 1966.
Extended Gaussian basis.

43. J.R. de la Vega, Y. Fang, and H.F. Hameka, *Physica, 's Grav.* **36**, 577 (1967).
Diamagnetic susceptibility.

44. V. Magnasco and A. Perico, *J. chem. Phys.* **47**, 971 (1967).
Localized orbitals.

45. G. Das and D.C. Wahl, *J. chem. Phys.* **47**, 2934 (1967).
Extended Hartree–Fock.

46. J.R. Hoyland, *J. chem. Phys.* **47**, 3556 (1967).
One centre.

47. A.D. McLean and H. Yoshimine, *J. chem. Phys.* **47**, 3256 (1967).
Rotational *g* factor, polarizabilities, magnetic properties.

48. Y. Kato and A. Saika, *J. chem. Phys.* **46**, 1975 (1967).
Nuclear spin–spin coupling.

49. J.M. O'Hare and R.P. Hurst, *J. chem. Phys.* **46**, 2356 (1967).
Hyperpolarizabilities.

50. P.E. Cade and W.M. Huo, *J. chem. Phys.* **47**, 614 (1967).
Hartree–Fock limit.

51. R.F.W. Bader, I. Keaveny and P.E. Cade, *J. chem. Phys.* **47**, 3381 (1967).
Binding characteristics.

52. B.J. Ransil and J.J. Sinai, *J. chem. Phys.* **46**, 4050 (1967).
Charge densities.

53. A.A. Frost, B.H. Prentice, and R.A. Rouse, *J. Am. chem. Soc.* **89**, 3064 (1969).
Floating spherical gaussian calculation.

54. J.R. de la Vega and H.F. Hameka, *J. chem. Phys.* **47**, 1834 (1967).
Magnetic properties.

55. S.R. La Paglia, *Theor. Chim. Acta* **8**, 185 (1967).
Transition probabilities.

56. P. Pyykko, *Proc. phys. Soc.* **92**, 841 (1967).
Quadrupole coupling constant.

57. A.D. McLean and H. Yoshimine, *J. chem. Phys.* **47**, 3256 (1967).
Properties calculated from Hartree–Fock function.

58. C.F. Bender and E.R. Davidson, *J. chem. Phys.* **47**, 360 (1967).
Natural orbitals.

59. F.E. Harris and H.H. Michaels, *Int. J. Quantum Chem.* **1S**, 329 (1967).
Double zeta basis.

60. R.M. Dixon and J.A.S. Smith, *Trans. Faraday Soc.* **64**, 1 (1968).
Deuteron coupling constant.

61. C.F. Bender and E.R. Davidson, *J. chem. Phys.* **49**, 4989 (1968).
Includes excited states.

62. M. Klessinger, *Chem. Phys. Lett.* **2**, 562 (1968).
Group functions.

63. A.C. Hopkinson, N.K. Holbrook, K. Yates and I.G. Csizmadia, *J. chem. Phys.* **49**, 3596 (1968).
Gaussian basis.

64. F. Grimaldi, A. Lecourt and C.M. Moser, *Symp. Faraday Soc.* **2**, 59 (1968).
Dipole moment.

65. C.F. Bender and E.R. Davidson, *Phys. Rev.* **183**, 23 (1969).
Natural orbitals + CI.

66. R.F.W. Bader and J.L. Ginsberg, *can J. Chem.* **47**, 3061 (1969).
Force constant.

67. H. Preuss and R. Janoschek, *J. molec. Struct.* **3**, 423 (1969).
Preliminary calculation, $E = -99{\cdot}728$.

FH^+

State	Internuclear distance	Basis set	Energy	Wave function	Reference
$X\,^2\Pi_i$	1·7328	Ext. STO	−99·53598		4
	1·8	Ext. STO	−99·5550		5
$^2\Sigma^+$	2·5	Ext. STO	−99·3988	✓	2
	2·0	Ext. STO	−99·4202		5

Properties	*References*
Spectroscopic constants	1,2,3,5
Excitation energies	3
Spin–orbit coupling constant	6

REFERENCES

1. S. Fraga and B.J. Ransil, *J. chem. Phys.* **35**, 669 (1961). Uses HF molecule orbitals.
2. E. Clementi, *J. chem. Phys.* **36**, 33 (1962). Uses HF molecule orbitals.
3. R.K. Nesbet, *J. chem. Phys.* **36**, 1518 (1962). Uses HF molecule orbitals.
4. P.E. Cade and W.M. Huo, *J. chem. Phys.* **47**, 649 (1967). Accurate Hartree–Fock result.
5. W.G. Richards and R.C. Wilson, *Trans. Faraday Soc.* **64**, 1729 (1968). Excited states.
6. T.E.H. Walker and W.G. Richards, *Symp. Faraday Soc.* **2**, 64 (1968). Spin–orbit coupling constant.

H_2

State	Internuclear distance	Basis set	Energy	Wave function	Reference
$X\,^1\Sigma_g^+$	1·474	Floating spherical Gaussian	−0·956		45
	1·4	One centre	−1·104		57
	1·4	GTO	−1·1266		55
	1·431	Weinbaum fns.	−1·14794		31
	1·402	Min. STO + CI	−1·15919	✓	10

State	Internuclear distance	Basis set	Energy	Wave function	Reference
	1·4	One centre + CI	−1·6042		20
	1·4	Different orbitals for different spin + CI	−1·16862		18
	1·4	Ext. HF Natural orbitals	−1·169837	√	36
	1·4	Elliptic fns.	−1·173128		
		James–Coolidge trial fn.	−1·1738		23
	$1{\cdot}0 < R < 3{\cdot}2$	100-term James–Coolidge fn. double precision	−1·17447498	√	54
$B\ ^1\Sigma_u^+$	2·3	Elliptic fns. + CI	−0·74371		7
	2·0	Elliptic fns. + CI	−0·75042		34
	2·43	Generalised James–Coolidge fn.	−0·7566611		37
$C\ ^1\Pi_u$	1·947	Elliptic fn. + CI	−0·7145		22
	2·0	Elliptic fn. + CI	−0·717832	√	34
	1·951	Generalized James–Coolidge	−0·7183492	√	54
$E\ ^1\Sigma_g^+$	1·9	Elliptic fns. + CI	−0·71643	√	27
	1·92	Elliptic fns.	−0·7179		9
		50-term wave fn.	−0·7181172		59
$F\ ^1\Sigma_g^+$		50-term wave fn.	−0·7145019		59
$3d\ ^1\Delta_g$	1·98	Elliptic fns. + CI	−0·6570		23
$a\ ^3\Sigma^+$	1·87	Elliptic fns.	−0·72730	√	27

State	Internuclear distance	Basis set	Energy	Wave function	Reference
	1·864	Elliptic fns.	−0·736615	√	29
	1·8683	James–Coolidge type	−0·7371522		41
$b\ ^3\Sigma_u^+$	Repulsive curve	Elliptic fns.			17, 34
	Repulsive curve	Generalized James–Coolidge		√	54
$c\ ^3\Pi_u$	1·96	Elliptic fns. + CI	−0·733439	√	23
	2·00	Elliptic fns. + CI	−0·736849	√	34
$^3\Pi_g$	1·990	One centre	−0·65817	√	35
$^3\Delta_g$	1·99	Elliptic fns. + CI	−0·6972	√	23
Many excited		Elliptic fns. + CI		√ √	33 34
		Bare nucleus purturbation theory			42

Properties	*References*
Dissociation Energy	1, 5, 16, 24, 25, 32, 37, 38, 56, 57, 60
Spin–spin and spin–orbit coupling	13
Magnetic properties	15
Spectroscopic constants	10, 14, 22, 31, 33, 34, 36, 46, 54
Polarizability	22, 43
Quadrupole moment	22, 57
Field gradients	44
Transition moments	62

REFERENCES

1. H.M. James and A.S. Coolidge, *Phys. Rev.* **43**, 588 (1933). D_e = 4·2 eV.

2. H.M. James and A.S. Coolidge, *J. chem. Phys.* **1**, 825 (1933). Correlated wave function.

3. H.M. James and A.S. Coolidge, *J. chem. Phys.* **6**, 730 (1938). Excited states.

4. F. Berencz, *Acta phys. hung.* **4**, 149 (1954).
Correlated natural orbitals.

5. F.E. Harris, *J. chem. Phys.* **27**, 812 (1957).
Survey of binding energies.

6. F. Berencz, *Acta phys. hung.* **6**, 423 (1957).
Correlated natural orbitals.

7. F. Berencz, *Acta phys. hung.* **10**, 389 (1959).
Review.

8. E.R. Davidson, *J. chem. Phys.* **33**, 1577 (1960).
Double minimum in excited state

9. E.R. Davidson, *J. chem. Phys.* **35**, 1189 (1961).
Excited $^1\Sigma_g^+$ state.

10. S. Fraga and B.J. Ransil, *J. chem. Phys.* **35**, 1967 (1961).
Min. basis.

11. E. R. Davidson and L.L. Jones, *J. chem. Phys.* **37**, 1918 (1962).
Correlation splitting.

12. G.M. Leies, *J. chem. Phys.* **37**, 1418 (1962).
Theoretical vibrational levels.

13. P. Fontana, *Phys. Rev.* **125**, 220 (1962).
Spin–spin and spin–orbit coupling.

14. D.M. Bishop, *Molec. Phys.* **6**, 305 (1963).
Single centre.

15. J.R. Hoyland and R.G. Parr, *J. chem. Phys.* **38**, 2991 (1963).
Magnetic properties.

16. W. Kolos and L. Wolniewicz, *Rev. mod. Phys.* **35**, 473 (1963).
Non-adiabatic four-particle calculation.

17. H.S. Taylor, *J. chem. Phys.* **39**, 3375 (1963).
Excited states.

18. E.G. Harris and H. Taylor, *J. chem. Phys.* **38**, 2591 (1963).
Different orbitals for different spins.

19. S. Hagstrum and H. Shull, *Rev. mod. Phys.* **35**, 624 (1963).
Natural orbitals.

20. D.M. Bishop, *Molec. Phys.* **6**, 305 (1963).
One centre.

21. M. Karplus and H.J. Kolker, *J. chem. Phys.* **39**, 2011 (1963).
Polarizability.

22. J.C. Browne, *J. chem. Phys.* **40**, 43 (1964).
Excited states

23. J. Goodisman, *J. chem. Phys.* **41**, 3889 (1964).
James and Coolidge function.

24. W. Kolos and L. Wolniewicz, *J. chem. Phys.* **41**, 3663 (1964).
Accurate adiabatic calculation.

25. W. Kolos and L. Wolniewicz, *J. chem. Phys.* **41**, 3674 (1964).
Accurate non-adiabatic calculation.

26. M.R. Flannery, *Proc. phys. Soc.* **85**, 1318 (1965).
Weinbaum function.

27. J. Gerhauser and H.S. Taylor, *J. chem. Phys.* **42**, 3261 (1965).
Excited states.

28. W.M. Wright and E.R. Davidson, *J. chem. Phys.* **43**, 840 (1965).
Excited state.

29. C. Wakefield and E.R. Davidson, *J. chem. Phys.* **43**, 834 (1965).
Excited state.

30. R.E. Christoffersen, *J. chem. Phys*, **42**, 2616 (1965).
Orbit–orbit interaction.

31. M.N. Adamov and V.P. Bulychev, *Teor. eksp. Khim.* **2**, 685 (1966).
Weinbaum function.

32. L. Wolniewicz, *J. chem. Phys.* **45**, 515 (1966).
Numerical integration.

33. S. Rothenberg and E.R. Davidson, *J. chem. Phys.* **44**, 730 (1966).
Several excited states.

34. S. Rothenberg and E.R. Davidson, *J. chem. Phys.* **45**, 2560 (1966).
Many excited levels.

35. J.R. Hoyland, *J. chem. Phys.* **45**, 3928 (1966).
One-centre calculation of excited states.

36. G. Das and A.C. Wahl, *J. chem. Phys.* **44**, 87 (1966).
Extended Hartree–Fock. Natural orbitals.

37. W. Kolos and L. Wolniewicz, *J. chem. Phys.* **45**, 509 (1966).
Generalized James and Coolidge function.

38. W. Kolos and L. Wolniewicz, *J. chem. Phys.* **45**, 944 (1966).
Dissociation energy.

39. J. Goodisman, *J. chem. Phys.* **47**, 1256 (1967).
Perturbation treatment.

40. Z. Dvoracek and Z. Harak, *J. chem. Phys.* **47**, 1211 (1967).
One-centre.

41. W. Kolos, *Chem. Phys. Lett.* **1**, 19 (1967).
Excited state.

42. C.A. Coulson and Z. Luz, *Mh. Chem.* **98**, 967 (1967).
Electron correlation.

43. W. Kolos and L. Wolniewicz, *J. chem. Phys.* **46**, 1426 (1967).
Polarizability

44. R.C. Henderson and D.D. Ebbing, *J. chem. Phys.* **47**, 69 (1967).
Field gradient from one-centre function.

45. A.A. Frost, B.H. Prentice and R.A. Rouse, *J. Am. chem. Soc.* **89**, 3064 (1969).
Floating spherical gaussian.

46. W. Kolos and L. Wolniewicz, *J. chem. Phys.* **48**, 3672 (1968).
Excited states.

47. B. Kirtman and D.R. Decius, *J. chem. Phys.* **48**, 3133 (1968).
Variation perturbation. Bare nucleus.

48. J. Goodisman, *J. chem. Phys.* **48**, 2981 (1968).
Bare nucleus perturbation theory.

49. R.L. Matcha and W. Byers-Brown, *J. chem. Phys.* **48**, 74 (1968).
Perturbation variation method.

50. Y.G. Smeyers, *An. R. Soc. esp. Fis. Quim.* **A64**, 263 (1968).
One-centre calculation.

51. R.J. Boys, *J. molec. Spectrosc.* **26**, 35 (1968).

52. J.N. Murrell and G. Shaw, *Theor. Chim. Acta*, **11**, 434 (1968).
Long-range forces.

53. W. Kolos, *Int. J. Quantum Chem.* **2**, 471 (1968).
Review of one- and two-electron molecules.

54. W. Kolos and L.Wolniewicz, *J. chem. Phys.* **49**, 404 (1968).
Double precision wave function.

55. A.C. Hopkinson, N.K. Holbrook, K. Yates, and I.G. Csizmadia *J. chem Phys.* **49**, 3596 (1968).
Gaussian basis set.

56. G. Berthier and H. Magot, *J. Chim. phys.* **56**, 504 (1969).
Elliptical coordinates.

57. J.A. Keefer, J.K. Su Zu and R.L. Belford, *J. chem. Phys.* **50**, 160 (1969).
One-centre calculation.

58. R.C. Morrison and G.A. Gallup, *J. chem. Phys.* **50**, 1214 (1969).
Unrestricted Hartree–Fock + CI.

59. W. Kolos and L. Wolniewicz, *J. chem. Phys.* **50**, 3228 (1969).
Excited states.

60. W. Kolos and L. Wolniewicz, *J. chem. Phys.* **51**, 1417 (1969).
Dissociation energy.

61. W.T. Zemke, P.G. Lykos, and A.C. Wahl, *J. chem. Phys.* **51**, 5635 (1969).
Excited states.

62. L. Wolniewicz, *J. chem. Phys.* **51**, 5082 (1969).

63. H.P. Kelly, *Phys. Rev. Lett.* **23**, 455 (1969).
Application of many-body perturbation theory.

H_2^+

The hydrogen-molecule ion is a very special case. As far as molecular wave functions are concerned the Schrödinger equation can be solved exactly for fixed nuclei. It is not appropriate to treat this molecule like all the others in the book as a comprehensive review of the calculations performed on H_2^+ would be a lengthy monograph in itself. Here we merely present a bibliography of some of the calculations on the ion which are in the same spirit as those for other molecules.

REFERENCES

1. D.R. Bates, R.K. Lidsham, and A.L. Stewart, *Phil. Trans. R. Soc.* **246**, 15 (1953).

2. H. Shull and D.B. Ebbing, *J. chem. Phys.* **28**, 866 (1958). Floating wave function.

3. E.M. Roberts, M.R. Foster, and F.F. Selig, *J. chem. Phys.* **37**, 485 (1962). Spin–orbit and hyperfine interactions.

4. B.P. Johnson and C.A. Coulson, *Proc. phys. Soc.* **84**, 263 (1964). Lower bound for energy.

5. S. Golden and J.G. Chernin, *J. chem. Phys.* **40**, 1032 (1964). Statistical theory. Dissociation energy.

6. J. Goodisman and D. Secrest, *J. chem. Phys.* **41**, 3610 (1964). Weinstein calculation for two-parameter James function.

7. K.C. Bhalla and P.G. Khubchandani, *Molec. Phys.* **9**, 229 (1965). M.O. with angularly dependent Z_{eff}.

8. K.C. Bhalla and P.G. Khubchandani, *Molec. Phys.* **9**, 291 (1965). Angularly dependent Z_{eff}. Dissociation energy.

9. H. Wind, *J. chem. Phys.* **42**, 2371 (1965). Long-range forces.

10. H.W. Joy and G.S. Handler, *J. chem. Phys.* **42**, 3047 (1965) One-centre expansion.

11. W.D. Lyon, R.L. Matcha, W.A. Saunder, W.J. Meath, and J.O. Hirschfelder, *J. chem. Phys.* **43**, 1095 (1965). Perturbation treatment.

12. S. Kim, T.Y. Chan, and J.O. Hirschfelder, *J. chem. Phys.* **43**, 1092 (1965). Guillemin–Zenner approximation.

13. W.B. Somerville, *J. chem. Phys.* **43**, 3398 (1965). Hyperfine interaction.

14. M. Walmsley and C.A. Coulson, *Proc. Camb. phil. Soc. math. phys. Sci.* **62**, 769 (1966). Lower bound to energy.

15. P.C. Haribaran and P.G. Khubchandani, *J. Phys.* B**1**, 134 (1967).
1st excited state.

16. K.C. Bhalla and P.G. Khubchandani, *Proc. phys. Soc.* **92**, 529 (1967).
Elliptical coordinates.

17. J. Lefaivre, *Can. J. Phys.* **45**, 228 (1967).
L.C.A.O. approximation.

18. J. Patel, *J. chem. Phys.* **47**, 770 (1967).
James and Coolidge function.

19. M.E. Schwartz and L.J. Schaad, *J. chem. Phys.* **46**, 4112 (1967).
Gaussian basis set.

20. J.V. Komarov and S.Y. Slavynov, *Zh. éksp. teor. Fiz.* **52**, 1368 (1967).

21. D. Iltena and H. Hartman, *Theor. Chim. Acta,* **7**, 1102 (1967).
Expectation values.

22. R.J. Damburg and R.K. Propius, *J. Phys.* B **1**, 681 (1968).
Large internuclear separations.

23. Y.C. Pan and H.F. Hameka, *J. chem. Phys.* **49**, 2009 (1968).
Analytical 2nd order energy using Greens function technique.

24. W. Kolos, *Int. J. Quantum Chem.* **2**, 471 (1968).
Review.

25. T.J. Houser, P.G. Lykos, and E.L. Hehler, *J. chem. Phys.* **38**, 583 (1958).
One-centre function.

26. D.M. Bishop, *J. chem. Phys.* **49**, 3718 (1968).
Electric field gradient.

27. R.G. Clark and E. T. Stewart, *J. Phys.* B **2**, 311 (1969).
3-parameter function.

28. S.K. Luke, G. Hunter, R.P. McEachran, and J.M. Cohen, *J. chem. Phys.* **50**, 1644 (1969).

29. D.M. Bishop and A. Macias, *J. chem. Phys.* **51**, 4997 (1969).
Force constant.

H_2^-

State	Internuclear distance	Basis set	Energy	Wave function	Reference
$^2\Sigma_u^+$	There is not complete agreement on whether or not this state is stable. Several references predict stability, others the converse.				1, 2, 3, 8 4, 5, 7

Properties	*References*
Dissociation energy	1, 2, 3
Potential curves	8

REFERENCES

1. I. Fischer-Hjalmars, *Ark. Fys.* **16**, 33 (1959).
   Predicts ground state and several excited states are stable.

2. I. Fischer-Hjalmars, *J. chem. Phys.* **30**, 1099 (1959).
   Dissociation energy.

3. B.K. Gupta, *Pnysica, 's Grav.* **25**, 190 (1959).

4. H.S. Taylor and F.E. Harris, *J. chem. Phys.* **39**, 1012 (1963).
   Suggests ground state unstable.

5. H.S. Taylor and J. Gerhauser, *J. chem. Phys.* **40**, 244 (1964).
   Suggests excited states are also unstable.

6. W.B. Somerville, *Proc. phys. Soc.* **89**, 185 (1966).
   Indicates importance of failure of adiabatic approximation.

7. H.S. Taylor, *Proc. phys. Soc.* **90**, 877 (1967).
   Against existence of the ion

8. P.G. Burke, *J. Phys.* B. **1**, 586 (1968).
   Potential curves.

He_2

State	Internuclear distance	Basis set	Energy	Wave function	Reference
$X\,^1\Sigma_g^+$	1·89	Min. STO	–5·544	✓	16
	1·89	GTO	–5·56530		34
	2·0	STO	–5·58131		9
	2·0	Ext. STO	–5·60253	✓	29
	1·89	Ext. STO	–5·60583	✓	17
	2·0	Elliptic	–5·6273	✓	32
	2·0	Scaled Atomic STO	–5·697		33
$^3\Sigma_u^+$	2·13	V.B., STO	–5·0938	✓	15
	2·139	V.B., ext. STO	–5·11346	✓	20
$^1\Sigma_u^+$	2·0	STO + elliptic	–5·063916	✓	21
	2·1	V.B., ext. STO	–5·10509	✓	23

State	Internuclear distance	Basis set	Energy	Wave function	Reference
$f\,^3\Pi_u$	2·0	V.B.	−4·97773	✓	22
	2·05	V.B.+CI	−5·03690		37
$F\,^1\Pi_u$	2·0	V.B.	−4·97755	✓	22
	2·05	V.B.+CI	−5·03649		37

Properties	*References*
Potential curve	4, 6, 7, 9, 10, 11, 13, 15, 16 17, 18, 19, 20, 21, 22, 23, 24, 25, 26, 27, 29, 30, 31, 32, 33, 37
Ionization potential	6
Charge density	7, 35
Overlap population	16
Orbital energies	16, 17, 29
Spectroscopic constants	20, 21, 22, 23, 37
Quadrupole moment	20

REFERENCES

1. J.C. Slater, *Phys. Rev.* **32**, 349 (1928).
Repulsive potential from V.B. using STO.

2. N. Rosen, *Phys. Rev.* **38**, 255 (1931).
Interaction energy at different *r* of 2 He atoms (normal) using V.B. with STO basis.

3. P. Rosen, *J. chem. Phys.* **18**, 1182 (1950).
Interaction potential of 2 He atoms.

4. R.A. Buckingham and A. Dalgarno, *Proc. R. Soc.* **A213**, 327 (1952).
Interaction of He (ground state) with He (excited state) using V.B. Wave functions from P.M. Morse, L.A. Young, and E.S. Haurwitz, *Phys. Rev.* **48**, 948 (1935).

5. H. Margenau and P. Rosen, *J. chem. Phys.* **21**, 394 (1952).
Interaction potential of 2 He atoms.

6. V. Griffing and J.F. Wehner, *J. chem. Phys.* **23**, 1024 (1955).
Roothaan method used on $X\,^1\Sigma_g^+$ with min. STO.

7. M. Sakamoto and E. Ishiguro, *Prog. theor. Phys., Osaka,* **15**, 37 (1956).
V.B. method using STO on $X\,^1\Sigma_g^+$.

8. S. Huzinaga, *Prog. theor. Phys., Osaka,* **17**, 169 (1957).
Repulsive potential from one-centre calculation on $X\,^1\Sigma_g^+$ using STO.

9. S. Huzinaga, *Prog. theor. Phys., Osaka,* **18**, 139 (1957).

10. S. Huzinaga, *Prog. theor. Phys., Osaka,* **20**, 15 (1958).
Single configuration M.O. study of $X\,^1\Sigma_g^+$ using STO.

11. T. Hashino and S. Huzinaga, *Prog. theor. Phys., Osaka*, **20**, 631 (1958).
Interaction energy of 2 He atoms using STO basis.

12. N. Lynn, *Proc. phys. Soc.* **A72**, 201 (1958).
Perturbation treatment of interaction energy of 2 He atoms.

13. N. Moore, *J. chem. Phys.* **33**, 471 (1960).
V.B. treatment of $X\,{}^1\Sigma_g^+$ using STO.

14. L. Salem, *Molec. Phys.* **3**, 441 (1960).
Perturbation treatment of interaction of 2 He atoms.

15. G.H. Brigman, S.J. Brient, and F.A. Matsen, *J. chem. Phys.* **34**, 958 (1961).

16. B.J. Ransil, *J. chem. Phys.* **34**, 2109 (1961).
Comparison with other calculations included.

17. P. Phillipson, *Phys. Rev.* **125**, 1981 (1962).
Single- and multi-configurational techniques employed.

18. D.Y. Kim, *Z. Phys.* **166**, 359 (1962).
Variational treatment of He/He interaction in ground state.

19. R.V. Miller and R.D. Present, *J. chem. Phys.* **38**, 1179 (1963).
One-centre treatment of $X\,{}^1\Sigma_g^+$ using STO + CI.
Wave functions given.

20. R.D. Poshusta and F.A. Matsen, *Phys. Rev.* **132**, 307 (1963).
Polarization included. Open and closed shell.wave functions used.

21. J.C. Browne, *J. chem. Phys.* **42**, 2826 (1965).
Also ${}^3\Sigma_g^+$ and ${}^1\Sigma_g^+$.

22. J.C. Browne, *Phys. Rev.* **138A**, 9 (1965).

23. D.R. Scott, E.M. Greenwalt, J.C. Browne, and F.A. Matsen, *J. chem. Phys.* **44**, 2981 (1966).

24. N.A. Kestner and O. Sinanoglu, *J. chem. Phys.* **45**, 194 (1966).
Pair-correlation method. Long-range interaction.

25. N.A. Kestner, *J. chem. Phys.* **45**, 208 (1966).
Pair-correlation method. Long-range interaction.

26. G.H. Matsumato, C.F. Bender, and E.R. Davidson, *J. chem. Phys.* **46**, 402 (1967).
Elliptical function in CI using natural orbitals to give energies at small r for $X\,{}^1\Sigma_g^+$.

27. B.M. Maris and R.D. Present, *J. chem. Phys.* **46**, 653 (1967).
One-centre CI calculation with ext. STO to give energies at small r for $X\,{}^1\Sigma_g^+$.

28. C.J. Herbert and O.G. Ludwig, *J. chem. Phys.* **47**, 3086 (1967).
Non integral quantum numbers used in repeat calculation of ref. 27.

29. T.L. Gilbert and A.C. Wahl, *J. chem Phys.* **47**, 3425 (1967).

30. D.J. Klein, C.E. Rodriguez, J. Browne, and F.A. Matsen, *J. chem.* **47**, 4862 (1967).
Elliptical and STO basis for $X\ ^1\Sigma_g^+$.

31. K. Kay and A.G. Turner, *Int. J. Quantum Chem.* **1S**, 167 (1967).
One-centre calculation using STO. Wave functions given.

32. G.P. Barnett., *Can. J. Phys.* **45**, 137 (1967).
Single- and multi-configurational techniques employed.

33. A.A. Wu and F.O. Ellison, *J. chem. Phys.* **48**, 1103 (1968).
'Atoms in molecules' method.

34. M.E. Schwartz and L.J. Schaad, *J. chem. Phys.* **48**, 4709 (1968).

35. R.F.W. Bader and A.K. Chandra, *Can. J. Chem.* **46**, 953 (1968).

36. J. Goodisman, *J. chem. Phys.* **50**, 903 (1969).
Perturbation treatment to obtain correlated wave functions.

37. B.K. Gupta and F.A. Matsen, *J. chem. Phys.* **50**, 3797 (1969).
STO for $^1\Pi_u$, STO + elliptical for $^3\Pi_u$.

He_2^+

State	Internuclear distance	Basis set	Energy	Wave function	Reference
$X\,^2\Sigma_u^+$	2·065	STO	4·890		5
	2·06	GTO	– 4·92109	✓	13
	2·0626	Ext. GTO	– 4·92159	✓	8
	2·0	Scaled atom STO	– 4·943		12
	2·0	Ext. STO	– 4·98540		11
	2·0625	Ext. STO	–4·98594	✓	7
	2·00		–4·9975		10
$^2\Sigma_g^+$	2·0	Scaled atom STO	–4·596		12
	2·0	Ext. STO	–4·59897		11

Properties	*References*
Spectroscopic constants	1, 2, 5, 7
Potential curve	1, 3, 7, 8, 9, 10, 11, 12, 13
Dissociation energy	3, 4, 6
Orbital energies	13

REFERENCES

1. L. Pauling, *J. chem. Phys.* **1**, 56 (1933).
Variational method using STO on $X\,^2\Sigma_u^+$.

2. S. Weinbaum, *J. chem. Phys.* **3**, 547 (1935).
Variational method using STO on $X\,^2\Sigma_u^+$.

3. B.L. Moisewitsch, *Proc. phys. Soc.* **A69**, 653 (1956).
Variational method using STO on $X\,^2\Sigma_u^+$ and $^2\Sigma_g^+$.

4. P. Czavinsky, *J. chem. Phys.* **31**, 178 (1959).
Variational method using STO on $X\,^2\Sigma_u^+$.

5. S. Fraga and B.J. Ransil, *J. chem. Phys.* **35**, 669 (1961).

6. F.A. Matsen and J.C. Browne, *J. phys. Chem., Ithaca*, **66**, 2332 (1962).
Wave functions (STO basis) given for $X\,^2\Sigma_u^+$.

7. P.N. Reagen, J.C. Browne, and F.A. Matsen, *Phys. Rev.* **B2**, 304 (1963).
Variational method.

8. C. Edmiston and M. Krauss, *J. chem. Phys.* **45**, 1833 (1966).
Superposition of configuration using pseudo-natural orbitals.

9. J.C. Browne, *J. chem. Phys.* **45**, 2707 (1966).
V.B. (STO and elliptic basis) on $^4\Sigma_u^+$, $^2\Sigma_g^+$, $^4\Sigma_g^+$ $(^2\Sigma_g^+)^*$, and $(^2\Sigma_u^+)^*$,

10. H. Conroy and B.L. Bruner, *J. chem. Phys.* **47**, 921 (1967).
See papers by H. Conroy for details of method and basis set.
J. chem. Phys. **41**, 1327, 1331, 1336, 1341 (1964).

11. B.K. Gupta and F.A. Matsen, *J. chem. Phys.* **47**, 4860 (1967).
Basis set given in ref. 7.

12. A.A. Wu and F.O. Ellison, *J. chem. Phys.* **48**, 1103 (1968).
Atoms in molecules method.

13. M.E. Schwartz and L.J. Schaad, *J. chem. Phys.* **48**, 4709 (1968).

He_2^{++}

State	Internuclear distance	Basis set	Energy	Wave function	Reference
$X\,^1\Sigma_g^+$	1·3	GTO	−3·60930	✓	6
	1·3	Ext. STO + CI	−3·66562	✓	3
	1·32	40-term James–Coolidge function	−3·6798		2
	1·4	STO	−3·681		7
	1·375		−3·681		5
$B\,^1\Sigma_u^+$	1·4	STO	−2·471 (not min.)		7
	3·25	STO + elliptical	−2·91472	✓	4

State	Internuclear distance	Basis set	Energy	Wave function	References
$b\ ^1\Sigma_u^+$	1·4	STO	−3·039 (not min.)		7
$E\ ^1\Sigma_g^+$	1·4	STO	−1·957 (not min.)		7
	3·5	STO + elliptical	−2·91279	✓	4

Properties	*References*
Spectroscopic constants	1, 3, 4
Potential curves	1, 2, 3, 4, 5, 6, 7

REFERENCES

1. L. Pauling, *J. chem. Phys.* **1**, 56 (1933).
Variational method using STO on $X\ ^1\Sigma_g^+$.
2. W. Kolos and C.C.J. Roothaan, *Rev. mod. Phys.* **32**, 219 (1960).
3. S. Fraga and B.J. Ransil, *J. chem. Phys.* **37**, 1112 (1962).
Single- and multi-configurational wave functions reported.
4. J.C. Browne, *J. chem. Phys.* **42**, 1428 (1965).
$^1\Sigma_u^+$, $^3\Sigma_u^+$, $^3\Sigma_g^+$, and $(^1\Sigma_g^+)^*$ discussed. Spectroscopic constants of bonding singlet states given. Potential curves of these and repulsive triplet states included.
5. H. Conroy and B.L. Bruner, *J. chem. Phys.* **47**, 921 (1967).
See papers by H. Conroy for details of method and basis set, i.e. *J. chem. Phys.* **41**, 1327, 1336, 1341 (1964).
6. M.E. Schwartz and L.J. Schaad, *J. chem. Phys.* **46**, 4112 (1967).
7. F.O. Ellison and A.A. Wu, *J. chem. Pnys.* **47**, 4408 (1967).
Scaled 'atoms in molecules' method.

HeF

State	Basis set	Energy	Reference
$^1\Sigma^+$	Valence bond Ext. GTO lobe	Repulsive potential curve	1

Properties	*Reference*
Potential curve	1

REFERENCE

1. L.C. Allen, A.M. Leek, and R.M. Erdahl, *J. Am. chem. Soc.* **88**, 615 (1966).
Finds repulsive potential curve.

HeH

State	Internuclear distance	Basis set	Energy	Wave function	Reference
$X\ ^2\Sigma^+$	3·0	Elliptic + CI	–3·35670	✓	2
	3·0	Elliptic + CI	–3·3618		3
	3·0	Elliptic, natural orbitals	–3·38280		4
$^2\Pi$	1·5	Elliptic + CI	–3·06983	✓	2
$^2\Sigma^+$	1·5	Elliptic + CI	–3·08304	✓	2

Properties	*References*
Dipole moment	4
Quadrupole moment	4
Spectroscopic constants	2

REFERENCES

1. A.C. Hurley, *Proc. phys. Soc.* **A69**, 868 (1956). Semi-empirical.
2. H.H. Michels and F.E. Harris, *J. chem. Phys.* **39**, 1464 (1963). Different orbitals for different spins.
3. H.S. Taylor and F.E. Harris, *Molec. Phys.* **7**, 287 (1964). Different orbitals for different spins.
4. C.F. Bender and E.R. Davidson, *J. phys. Chem., Ithaca* **70**, 2675 (1966). Natural orbitals.
5. A.A. Wu and F.O. Ellison, *J. chem. Phys.* **48**, 1103 (1963). Atoms in molecules method.

HeH^+

State	Internuclear distance	Basis set	Energy	Wave function	Reference
$X\ ^1\Sigma^+$	1·40	One electron diatomic	–2·903		14
	1·4	GTO	–2·919		23
	1·48	One centre	–2·919		22

State	Internuclear distance	Basis set	Energy	Wave function	Reference
	1·40	GTO	−2·93943		16
	1·40	Ext. STO	−2·93259	✓	9
	1·40	Elliptic + CI	−2·94321		5
	1·40	Elliptic + CI	−2·94321	✓	13
	1·442	V.B. STO	−2·94920	✓	6
	1·50	Elliptic	−2·9550	✓	12
	1·40	Non integral Elliptic	−2·97190	✓	15
	1·40	James–Coolidge	−2·9727	✓	11
	1·446	One centre + CI	−2·9742	✓	8
	1·40	Elliptic + CI	−2·97424	✓	4
	1·4	One centre	−2·97458		18
	1·4		−2·97753 (Upper bound)	✓	7
	1·463	64-term James–Coolidge	−2·978669	✓	10
$A\,^1\Sigma^+$	Repulsive	James–Coolidge		✓	11, 10
	6·0	Elliptic + CI	−2·50124	✓	13
$B\,^1\Sigma^+$	8·5	Elliptic + CI	−2·18177	✓	13
$C\,^1\Pi$	8·0	Elliptic + CI	−2·13318	✓	13
	8·05	Elliptic and One centre	−2·133599	✓	17
$a\,^3\Sigma^+$	4·5	Elliptic + CI	−2·50300	✓	13
$b\,^3\Sigma^+$	8·0	Elliptic + CI	−2·20072	✓	13
$c\,^3\Pi$	8·0	Elliptic + CI	−2·13761	✓	13
	7·67	Elliptic and One centre	−2·138864	✓	17

Properties	*References*
Dipole moment	4, 9

Properties	*References*
Spectroscopic constant	1,4,5,8,9,10,11,12,17
Excitation energies	10,11,13,17
Potential curves	3,6,8,9,10,11
Charge density	9

REFERENCES

1. J.Y. Beach, *J. chem. Phys.* **4**, 353 (1936).

2. A.A. Evett, *J. chem. Phys.* **23**, 1169 (1955).
   James–Coolidge function.

3. J. Ross and E.A. Mason, *Astrophys. J.* **124**, 485 (1956).
   Long-range forces.

4. B.G. Anex, *J. chem. Phys.* **38**, 1651 (1963).

5. H.H. Michels and F.E. Harris, *J. chem. Phys.* **39**, 1464 (1963).

6. H. Preuss, *Molec. Phys.* **8**, 233 (1964).
   Gives potential curves.

7. H. Conroy, *J. chem. Phys.* **41**, 1341 (1964).
   Minimizes energy variance.

8. J.D. Stuart and F.A. Matsen, *J. chem. Phys.* **41**, 1646 (1964).
   One centre.

9. S. Peyerimhoff, *J. chem. Phys.* **43**, 998 (1965).

10. L. Wolniewicz, *J. chem. Phys.* **43**, 1087 (1965).
    64-term generalized James–Coolidge function.

11. J. Goodisman, *J. chem. Phys.* **43**, 3037 (1965).
    James–Coolidge function.

12. F.E. Harris, *J. chem. Phys.* **44**, 3636 (1966).
    Elliptic function basis.

13. H.H. Michels, *J. chem. Phys.* **44**, 3834 (1966).
    Several excited states.

14. G.A. Gallup and M.S. McKnight, *J. chem. Phys.* **45**, 365 (1966).
    Basis of exact one-electron diatomic functions.

15. J.R. Hoyland, *J. chem. Phys.* **45**, 466 (1966).

16. M.E. Schwartz and L.J. Schaad, *J. chem. Phys.* **46**, 4112 (1967).
    Gaussian basis.

17. J.R. Hoyland, *J. chem. Phys.* **47**, 49 (1967).
    Excited states.

18. F. Grein and T.J. Tseng, *Theor. Chim. Acta* **12**, 57 (1968).
    One centre.

19. K.E. Banyard and M.R. Hayns, *Molec. Phys.* **15**, 615 (1968).
    One centre.

20. L.L. Combs and L.K. Runnels, *J. chem. Phys.* **49**, 4126 (1968).
One centre.

21. A.A. Wu and F.O. Ellison, *J. chem. Phys.* **48**, 1103 (1968).
Atoms in molecules approach.

22. J.A. Keefer, J.K. SuFu, and R.L. Belford, *J. chem. Phys.* **50**, 160 (1969).
One centre.

23. H. Preuss and R. Janoschek, *J. molec. Struct.* **3**, 423 (1969).

24. L. Piela, *Int. J. Quantum Chem.* **3**, 945 (1969).
Long-range interaction.

HeH^{++}

This is a one-electron molecule for which the Schrödinger equation can be solved exactly for fixed nuclei.

REFERENCES

1. D.R. Bates and T.R. Carson, *Proc. R. Soc.* **A234** 207 (1956).
$1s\sigma$, $2s\sigma$, $2p\sigma$, $2p\pi$, and $3d\sigma$ states. $2p\sigma$ appears to be bound.

2. A.A. Evett, *J. chem. Phys.* **24**, 150 (1956).
James–Coolidge function

3. M. Cohen, R.P. McEachran, and S.D. McPhee, *Can. J. Phys.* **45**, 2231 (1967).
$1s\sigma$ and $2p\pi$ states by perturbation theory.

4. L.L. Combs and L.K. Runnels, *J. chem. Phys.* **49**, 4216 (1968).
One centre.

HeLi

State	Basis set	Energy	Wave function	Reference
$X\,^2\Sigma^+$	Min. STO	repulsive	✓	1
	Elliptic + CI	repulsive	✓	2
Also many excited states, both doublets and quartets				1, 2

Properties	*References*
Potential curves	1, 2, 3
Spectroscopic constants	2, 3
Excited states	2
Term values	2

REFERENCES

1. N. Scheel and V. Griffing, *J. chem. Phys.* **36**, 1453 (1962). All states calculated to be repulsive.

2. S.B. Schneiderman and H.H. Michels, *J. chem. Phys.* **42**, 3706 (1965). Many excited states including some calculated to be bonding.

3. M. Krauss, *J. Res. natn. Bur. Stand.* **72A**, 553 (1968).

$HeLi^+$, $HeLi^-$

State	Basis set	Wave function	Reference
$X\ {}^1\Sigma^+$ and other states	Min. STO	✓	1
	Elliptic + CI	✓	2

Property	*References*
Potential curves	1, 2

REFERENCES

1. N. Scheel and V. Griffing, *J. chem. Phys.* **36**, 1453 (1962). Min. STO.

2. S.B. Schneiderman and H.H. Michels, *J. chem. Phys.* **42**, 3706 (1965). Elliptical basis set + CI.

HeO

State	Basis set	Wave function	Reference
${}^1\Sigma^+$	Ext. GTO lobe Valence bond	Repulsive potential curve	1

Property	*Reference*
Potential curve	1

REFERENCES

1. L.C. Allen, A.M. Lesk, and R.M. Erdahl, *J. Am. chem. Soc.* **88**, 615 (1966). Finds repulsive potential curve.

KCl

State	Internuclear distance	Basis set	Energy	Wave function	Reference
$^1\Sigma^+$	5·039	Ext. STO	−1058·7583	✓	1

Properties	*Reference*
Orbital energies	1
Potential curve	1

REFERENCE

1. A.D. McLean and M. Yoshimine, Supp. to *I.B.M. Jl Res. Dev.* (1967). Near Hartree–Fock limit.

KF

State	Internuclear distance	Basis set	Energy	Wave function	Reference
$^1\Sigma^+$	4·10348	Ext. STO	−698·6850	✓	1
	4·188	Ext. STO	−698·68501	✓	2

Properties	*References*
Orbital energies	1, 2
Potential curve	1, 2
Spectroscopic constants	2
Quadrupole constant	2

REFERENCES

1. A.D. McLean and M. Yoshimine, Supp. to *I.B.M. Jl Res. Dev.* (1967). Very near Hartree–Fock limit.

2. R.L. Matcha, *J. chem. Phys.* **49**, 1264 (1968). Very near Hartree–Fock limit. Many one-electron properties.

Li_2

State	Internuclear distance	Basis set	Energy	Wave function	Reference
$X\,^1\Sigma_g^+$	5·31	Floating spherical gaussian	−12·282		37
	5·0	Min. STO	−14·8177	✓	5

State	Internuclear distance	Basis set	Energy	Wave function	Reference
	5·051	Min. STO	−14·84149	✓	14
	5·34	Min. STO	−14·84220		18
	5·051	Min.STO +CI	−14·8523	✓	20
	5·0	V.B.	−14·8635		11
	5·0	STO+CI	−14·8652		14
	5·051	Min. STO	−14·87152		32
	5·0	Double zeta basis	−14·87285		33
	5·25	Extended Hartree–Fock			27
		Valence with Ext. STO	−14·899	✓	27
	5·07	Extended Hartree–Fock Valence with Ext. STO	−14·90260		30
$^1\Sigma_u^+$	5·76	Min. STO	−14·75376		18
	5·0	V.B.	−14·7912		11
	5·0	STO+CI	−14·7946		14
$^1\Sigma_g^+$	5·88	Min. STO	−14·70738		18
$^3\Sigma_g^+$	5·74	Min. STO	−14·777287		18
$^3\Sigma_u^+$	5·0	V.B.	−14·8145		11
$^1\Pi_u$	5·0	V.B.	−14·7409		11
	5·0	STO+CI	−14·7452		14

Properties	*References*
Potential curve	1, 11, 13, 18, 27, 30, 33
Dissociation energy	2, 4, 13, 29
Field gradient	6, 7, 11, 15
Nuclear quadrupole moment	7, 8, 10, 11
Term values	9, 13, 18
Ionization potential	10
Population analysis	10, 11, 16
Spectroscopic constants	11, 13, 18, 27, 30
Nuclear quadrupole coupling	12, 19
Orbital energies	14
Core polarization	17
Magnetic properties	21, 22, 25, 26, 35, 36

Properties	*References*
Electric polarizability	23, 31
Charge density maps	28, 29, 39
Orbital forces	29

REFERENCES

1. M. Delbrück, *Annln Phys.* **5**, 36 (1930).
   Ground-state potential curve from V.B. calculation
2. W.H. Furry and J.H. Bartlett, Jr., *Phys. Rev.* **38**, 1615 (1931).
   V.B. calculation on ground state.
3. W.H. Furry and J.H. Bartlett, Jr., *Phys. Rev.* **39**, 1015 (1932).
   Phys. Rev. **43**, 361 (1933).
   $^1\Sigma$, $^1\Pi$ excited states treated by V.B.
4. H.M. James, *J. chem. Phys.* **2**, 794 (1934).
   V.B. Calculation on ground state.
5. C.A. Coulson and W.E. Duncanson, *Proc. R. Soc.* **A181**, 378 (1943).
6. P. Kusch, *Phys. Rev.* **76, 138 (1949)**.
   Reference to calculation by H.M. Foley on electric field gradient using wave functions given in ref. 2.
7. E.G. Harris and M.A. Melkanoff, *Phys. Rev.* **90**, 585 (1953).
   Repeat of claculation in ref. 6 but using wave function from ref. 4.
8. R.M. Sternheimer and H.M. Foley, *Phys. Rev.* **92**, 1460 (1953).
   Calculation use wave functions from ref. 4 and 5.
9. A. Rahman, *Physica, 's Grav.* **20**, 623 (1954).
   Calculations on $X\,^1\Sigma_g^+$, $^1\Sigma_u^+$, $^1\Pi_u$ using ASMO + CI and 'atoms in molecules' methods.
10. J.F. Faulkner, *J. chem. Phys.* **27**, 369 (1957).
    Calculation on $X\,^1\Sigma^+$ state using min. STO.
11. E. Ishiguro, K. Kayama, M. Kotani, and Y. Mizimo, *J. phys. Soc. Japan* **12**, 1355 (1957).
    Superposition of configuration V.B. calculation. Other methods also included.
12. I. Marmari and T. Arai, *J. chem. Phys.* **28**, 28 (1958).
    Method of 'deformed atoms in molecules' used on ground state with STO basis.
13. T. Arai and M. Sakamoto, *J. chem. Phys.* **28**, 32 (1958).
    $X\,^1\Sigma_g^+$, $^3\Sigma_u^+$, $^1\Sigma_u^+$, $^1\Pi_u$ states treated by method of 'deformed atoms in molecules' using STO basis.
14. B.J. Ransil, *Rev. mod. Phys.* **32**, 239, 245 (1960).
15. J.W. Richardson, *Rev. mod. Phys.* **32**, 461 (1960).
    Wave function in ref. 14 used.

16. B.J. Ransil and S. Fraga, *J. chem. Phys.* **34**, 727 (1961).
Wave functions in ref. 14 used.

17. O. Sinanoglu and E.M. Mortensen, *J. chem. Phys.* **34**, 1078 (1961).
2s Slater orbitals used.

18. S. Fraga and B.J. Ransil, *J. chem. Phys.* **35**, 669 (1961).
Wave function in ref. 14 used.

19. S.L. Kahalas and R.K. Nesbet, *Phys. Rev. Lett.* **6**, 549 (1961).
Ext. STO basis.

20. S. Fraga and B.J. Ransil, *J. chem. Phys.* **36**, 1127 (1962).

21. C.W. Kern and W.N. Lipscomb, *J. chem. Phys.* **37**, 260 (1962).
Wave functions in ref. 14 used.

22. M Karplus and H.J. Kolker, *J. chem. Phys.* **38**, 1263 (1963).
Variation, perturbation calculation using wave function in ref. 14.

23. M. Karplus and H.J. Kolker, *J. chem. Phys.* **39**, 2011 (1963).
Perturbation calculation using wave functions in ref. 14.

24. C. Manneback, *Physica, 's Grav.* **29**, 769 (1963).
CI coefficients given.

25. H.J. Kolker and M. Karplus, *J. chem. Phys.* **41**, 1259 (1964).
Variation, perturbation calculation using wave functions in ref. 14.

26. R.M. Stevens and W.N. Lipscomb, *J. chem. Phys.* **42**, 4302 (1965).
Calculation uses wave function in ref. 14 and some due to A.C. Wahl (unpub.).

27. G. Das and A.C. Wahl, *J. chem. Phys.* **44**, 87 (1966).

28. A.C. Wahl, *Science, N.Y.* **151**, 961 (1966).
STO basis.

29. R.F.W. Bader, W.H. Henneker, and P.E. Cade, *J. chem. Phys.* **46**, 3341 (1967).
STO Basis set due to P.E. Cade, K.D. Sales, and A.C. Wahl (unpub.).

30. G. Das, *J. chem. Phys.* **46**, 1568 (1967).
Configuration mixing method for diatomics described. Vectors and mixing coefficients for Li_2 given.

31. J.M. O'Hare and R.P. Hurst, *J. chem. Phys.* **46**, 2356 (1927).
Perturbation method.

32. B.J. Ransil and J.J. Sinai, *J. chem. Phys.* **46**, 4050 (1967).
Wave functions in ref. 14 and some due to A.C. Wahl (unpub.).

33. D.K. Rai and J.L. Calais, *J. chem. Phys.* **47**, 906 (1967).
Alternant MO calculation. Special case of different orbitals for different spins.

34. V. Magnasco and A. Perico, *J. chem. Phys.* **47**, 971 (1967).
Localization of atomic and molecular orbitals. Wave functions in ref. 14.

35. J.R. de la Vega and H.F. Hameka, *J. chem. Phys.* **47**, 1834 (1967)
Wave functions in ref. 14.

36. J.R de la Vega and H.F. Hameka, *Physica, s' Grav.* **35**, 313 (1967).
Wave functions in ref. 14.

37. A.A. Frost, B.H. Prentice, and R.A. Rouse, *J. Am. chem. Soc.* **89**, 3064 (1967).

38. M.A. Marchetti and S.R. La Paglia, *J. chem. Phys.* **48**, 434 (1968).
$^1\Sigma_u^+ - {}^1\Sigma_g^+$ dipole strengths from STO + CI treatment using STO basis in ref. 14.

39. R.F.W. Bader and A.K. Chandra, *Can. J. Chem.* **46**, 953 (1968).
Wavefunctions due to G. Das and A.C. Wahl. (see refs. 27 and 30).

Li_2^+

State	Internuclear distance	Basis set	Energy	Wave function	Reference
$X\,^2\Sigma_g^+$	3·432	Min. STO	−14·66922		2

Properties	*References*
Potential curves	2
Spectroscopic constants	2
Term values	2
Dissociation energy	1

REFERENCES

1. H.M. James, *J. chem. Phys.* **3**, 9 (1935).
Variational calculation (using STO) on ground state.

2. S. Fraga and B.J. Ransil, *J. chem. Phys.* **35**, 669 (1961).

LiBr

State	Internuclear distance	Basis set	Energy	Wave function	Reference
$X\,^1\Sigma^+$	4·0655	Ext. STO	−2579·8901	✓	1

Properties	*References*
Orbital energies	1
Potential curve	1

REFERENCE

1. A.D. McLean and M. Yoshimine, Supp. to *I.B.M. Jl Res. Dev.* (1967).
Near Hartree–Fock limit.

LiCl

State	Internuclear distance	Basis set	Energy	Wave function	Reference
$^1\Sigma^+$	3·825	Ext. STO	−457·05466	✓	1
	3·825	Ext. STO	−457·0547	✓	2

Properties	*References*
Potential curve	1, 2
Orbital energies	1, 2
Spectroscopic constants	1, 3
Dipole moment	1, 3
Quadrupole constant	1

REFERENCES

1. R.L. Matcha, *J. chem. Phys.* **47**, 4595 (1967).
Very near Hartree–Fock limit. Extensive range of one-electron properties.

2. A.D. McLean and M. Yoshimine, Supp. to *I.B.M. Jl Res. Dev.* (1961).
Very near Hartree–Fock limit.

3. H. Preuss and R. Janoschek, *J. molec. Struct.* **3**, 423 (1969).

LiF

State	Internuclear distance	Basis set	Energy	Wave function	Reference
$^1\Sigma^+$	2·85	Min. STO	−106·38126		1
		Min. STO	−106·40827	✓	3
	2·95536	GTO/V.B.	−106·5641		19
	2·89	Ext. STO	−106·97690		8
	2·8877	Ext. STO	−106·98850	✓	7
	2·8877	Ext. STO	−106·9916		11
	2·8877	Ext. STO	−106·9916	✓	16

Properties	*References*
Dipole moment	1, 3, 7, 8, 11
Orbital energies	1, 16
Atomic population	2, 3
Magnetic shielding	4, 9
Electric polarizability	6
Quadrupole coupling	7
Magnetic susceptibility	5, 15, 17
Spectroscopic constants	9
Charge density	10, 13
Transition probabilities	12
Localized orbitals	14
Potential curves	16, 19
Electron distribution	18

REFERENCES

1. B.J. Ransil, *Rev. mod. Phys.* **32**, 239 (1960).
Min. STO.

2. B.J. Ransil, *J. chem. Phys.* **34**, 727 (1961).
Atomic population.

3. S. Fraga and B.J. Ransil, *J. chem. Phys.* **36**, 1127 (1962).
Limited CI also, with min. STO.

4. C.W. Kern and W.N. Lipscomb, *J, chem. Phys.* **37**, 260 (1962).
Magnetic shielding.

5. M. Karplus and H.J. Kolker, *J. chem. Phys.* **38**, 1263 (1963).
Magnetic susceptibility.

6. M. Karplus and H.J. Kolker, *J. chem. Phys.* **39**, 2011 (1963).
Electric polarizability.

7. A.D. McLean, *J. chem. Phys.* **39**, 2653 (1963).
Compares different basis sets.

8. A.D. McLean, *J. chem. Phys.* **40**, 2774 (1964).
Scale optimization employed.

9. H.J. Kolker and M. Karplus, *J. chem. Phys.* **41**, 1259 (1964).
Nuclear magnetic shielding.

10. R.F.W. Bader and W.H. Henneker, *J. Am. chem. Soc.* **87**, 3063 (1965)
Charge density.

11. M. Yoshimine and A.D. McLean, *Int. J. Quantum Chem.* **1S**, 313 (1967).
Very near Hartree–Fock limit.

12. S.R. La Paglia, *Theor. Chim. Acta* **8**, 185 (1967).
Transition probabilities.

13. B.J. Ransil and J.J. Sinai, *J. chem. Phys.* **46**, 4050 (1967). Charge density.

14. V. Magnasco and A. Perico, *J. chem. Phys.* **47**, 971 (1967). Localized orbitals.

15. J.R. de la Vega, D. Ziobro, and H.F. Hameka, *Physica, 's Grav.* **37**, 265 (1967). Magnetic susceptibility.

16. A.D. McLean and M. Yoshimine, Supp. to *I.B.M. Jl Res. Dev.* (1967). Very near Hartree–Fock limit.

17. J.R. de la Vega and H.F. Hameka, *J. chem. Phys.* **47**, 1834 (1967). Rotational magnetic moment, diamagnetic susceptibility.

18. W.H. Henneker and P.E. Cade, *Chem. Phys. Lett.* **2**, 575 (1968). Electron distribution in momentum space.

19. C.G. Balint-Kurti and M. Karplus, *J. chem. Phys.* **50**, 478 (1969). Valence bond. Correlation energy also estimated.

20. H. Preuss and R. Janoschek, *J. molec. Struct.* **3**, 423 (1969). Preliminary calculation, $E = -105{\cdot}9106$.

LiH

State	Internuclear distance	Basis set	Energy	Wave function	Reference
$X\,^1\Sigma^+$	3·0	Min. STO	−7·96992	✓	13, 14
	3·0	STO	−7·9758		10
	3·015	STO	−7·9792		71
	3·015	Min. STO + CI	−7·9836	✓	25
	3·02	Ext. GTO	−7·9842		50
	3·06	Floating GTO	−7·9851		82
	3·0581	Ext. STO (+ CI)	−7·98597	✓	33
	3·015	Ext. STO	−7·98731		64
	3·015	Valence bond	−8·017		81
	3·035	STO + CI	−8·01735	✓	84
	3·0	Germinal from elliptic basis	−8·0179		44
	3·0	Ext. GTO + CI	−8·0188	✓	72
	3·0524	V.B. + CI	−8·022309	✓	83

State	Internuclear distance	Basis set	Energy	Wave function	Reference
	3·2	Elliptic different orbitals for different spins	−8·0387	✓	36
	3·0	53-term function	−8·04127		27
	3·02	Correlated function	−8·044		85
	3·606	Elliptic + CI	−8·0556		74
	3·0147	45 config. NO	−8·0606	✓	54
	3·015	939 config.	−8·060		76
		soln. of transcorrelated wave equation	8·063	✓	78
$A\ ^1\Sigma^+$	4·928	Elliptic + CI	−7·9372	✓	74
	3·0	Group orbitals	−7·8773	✓	42
	4·90	Numerical Hartree–Fock	−7·8963		9
$^3\Sigma^+$		Group orbitals	Repulsive	✓	42
		Elliptic fns.	Repulsive		34
		Elliptic +CI	Repulsive		74

Properties	*References*
Dipole moment	13, 15, 17, 18, 20, 25, 33, 47, 50, 54, 55, 56, 75, 76
Spectroscopic constants	13, 41, 58, 67, 74, 76, 79, 84
Charge densities	11, 15, 19, 42, 52, 53, 68
Magnetic properties	21, 22, 30, 31, 32, 38, 40, 48, 57, 59, 60, 63, 69, 76
Electric field gradient	23, 37, 43, 77
Deuteron quadrupole coupling	26
Polarizability	35, 61
Quadrupole moment	54, 55

REFERENCES

1. E. Hutchinson and M. Muskat, *Phys. Rev.* **40**, 340 (1932).
   Crude V.B. calculation.

2. J.K. Knipp, *J. chem. Phys.* **4**, 300 (1935).
   Extremely good energy.

3. I. Fischer, *Ark. Fys.* **5**, 349 (1952).
   Min. basis. Discussion of approximations.

4. J.R. Miller, R.H. Friedman, R.P. Hurst, and F.A. Matsen, *J. chem. Phys.* **27**, 1385 (1957).

5. R.P. Hurst, J.R. Miller, and F.A. Matsen, *J. chem. Phys.* **26**, 1092 (1957).

6. J.R. Miller, R.H. Friedman, R.P. Hurst, and F.A. Matsen, *J. chem. Phys.* **27**, 1385 (1957).

7. F.T. Ormand, F.A. Matsen, *J. chem. Phys.* **29**, 100 (1958).

8. O. Platas and F.A. Matsen, *J. chem. Phys.* **29**, 965 (1958).
   Refs. 4–8 are valence bond calculations

9. A.M. Karo and A.R. Olsen, *J. chem. Phys.* **30**, 1232 (1959).
   Valence bond – 6 configurations.

10. A.M. Karo, *J. chem Phys.* **30**, 1241 (1959).
    Natural orbitals with and without CI.

11. A.M. Karo, *J. chem. Phys.* **31**, 182 (1959).
    Population analysis.

12. S. Basu, *C.r. hebd. Séanc. Acad. Sci., Paris* **249**, 689 (1959).
    Simplified calculation.

13. B.J. Ransil, *Rev. mod. Phys.* **32**, 239 (1960).
    Minimum basis.

14. B.J. Ransil, *Rev. mod. Phys.* **32**, 245 (1960).
    Gives wave functions of ref. 13.

15. A.M. Karo, *J. chem. Phys.* **32**, 907 (1960).
    Population analysis.

16. J.M. Robinson, J.D. Stuart, and F.A. Matsen, *J. chem. Phys.* **32**, 988 (1960).
    Radially correlated V.B. function.

17. R. Moccia, *Gazz. chim. ital.* **90**, 955 (1960).
    Min. basis.

18. R. Moccia, *Gazz. chim. ital.* **90**, 968 (1960).
    Min. basis.

19. S. Fraga and B.J. Ransil, *J. chem. Phys.* **34**, 727 (1961).
    Population analysis.

20. S. Fraga and B.J. Ransil, *J. chem. Phys.* **35**, 669 (1961). Includes calculation of excited states by virtual orbitals.

21. C.W. Kern and W.N. Lipscomb, *Phys. Rev. Lett.* **7**, 19 (1961). Magnetic shielding constants.

22. M. Karplus and H.J. Kolker, *J. chem. Phys.* **35**, 2235 (1961). Magnetic interactions.

23. S.L. Kahalas and R.K. Nesbet, *Phys. Rev. Lett.* **6**, 549 (1961). Electric field gradient.

24. F.O. Ellison, *J. phys. Chem. Ithaca* **66**, 2294 (1962). Semi-empirical.

25. S. Fraga and B.J. Ransil, *J. chem. Phys.* **36**, 1127 (1962). CI.

26. H.J. Kolker and M. Karplus, *J. chem. Phys.* **36**, 960 (1962). Deuteron quadrupole coupling.

27. D.D. Ebbing, *J. chem. Phys.* **36**, 1361 (1962). Extensive CI.

28. J.C. Browne, *J. chem. Phys.* **36**, 1814 (1962). Valence bond + CI.

29. F.A. Matsen and J.C. Browne, *J. phys. Chem., Ithaca* **66**, 2332 Review.

30. C.W. Kern and W.N. Lipscomb, *J. chem. Phys.* **37**, 260 (1962). Magnetic shielding.

31. R.M. Stevens, R.M. Pitzer, and W.N. Lipscomb, *J. chem. Phys.* **38**, 550 (1963). Perturbed Hartree–Fock.

32. M. Karplus and H.J. Kolker, *J. chem. Phys.* **38**, 1263 (1963). Magnetic susceptibility.

33. S.L. Kahalas and R.K. Nesbet, *J. chem. Phys.* **39**, 529 (1963). Extended basis set.

34. H.S. Taylor, *J. chem. Phys.* **39**, 3382 (1963). Excited state potential curve.

35. H.J. Kolker and M. Karplus, *J. chem. Phys.* **39**, 2011 (1963). Polarizability.

36. F.E. Harris and H.S. Taylor, *Physica, 's Grav.* **30**, 105 (1964). Different orbitals for different spins.

37. J.C. Browne and F.A. Matsen, *Phys. Rev.* **135**, A1227 (1964). Valence bond.

38. J.R. de la Vega and H.F. Hameka, *J. chem. Phys.* **40**, 1929 (1964). Magnetic susceptibility.

39. R.M. Stevens and W.N. Lipscomb, *J. chem. Phys.* **40**, 2238 (1964). Perturbed Hartree–Fock.

40. H.J. Kolker and M. Karplus, *J. chem. Phys.* **41**, 1259 (1964).
Nuclear magnetic shielding.

41. J.C. Browne, *J. chem. Phys.* **41**, 3495 (1964).
Ionization potential.

42. I.G. Csizmadia, B.T. Sutcliffe, and H.P. Barnett, *Can. J. Chem.* **42**, 1645 (1964).
Density contours.

43. Y. Rasiel and R.D. Whitman, *J. chem. Phys.* **42**, 2124 (1965).
Electric moments.

44. D.D. Ebbing and R.C. Henderson, *J. chem. Phys.* **42**, 2225 (1965).
Geminal expansion.

45. G.P. Barnett, J. Linderberg, and H. Shull, *J. chem. Phys.* **43**, S80 (1965).
Natural orbitals.

46. J.D. Stuart and R.P. Hurst, *Molec. Phys.* **9**, 265 (1965).
Valence electron calculation

47. P.E. Cade and W.M. Huo, *J. chem. Phys.* **45**, 1063 (1966).
Dipole moment.

48. G.P. Arrighini, F. Grossi, and M. Maestro, *Theor. Chim. Acta* **5**, 266 (1966).
Magnetic properties.

49. D.P. Chong and Y. Rasiel, *J. chem. Phys.* **44**, 1819 (1966).
Constrained variational method.

50. I.G. Csizmadia, *J. chem. Phys.* **44**, 1849 (1966).
Gaussian basis.

51. D.P. Chong and W.B. Brown, *J. chem. Phys.* **45**, 392 (1966).
Perturbation theory of constraints.

52. R.F.W. Bader and W.H. Henneker, *J. Am. chem. Soc.* **88**, 280 (1966).
Charge densities.

53. P. Politzer and R.E. Brown, *J. chem. Phys.* **45**, 451 (1966).
Density contours.

54. C.F. Bender and E.R. Davidson, *J. Phys. Chem., Ithaca* **70**, 2675 (1966).
54 configuration function.

55. D.E. Stogryn and A.F. Stogryn, *Molec. Phys.* **11**, 371 (1966).
Dipole and quadrupole moments.

56. P. Linder, *Theor. Chim. Acta,* **5**, 336 (1966).
Dipole moment.

57. J.R. de la Vega, and H.F. Hameka, *J. chem. Phys.* **47**, 1834 (1967).
Magnetic properties.

58. S.R. La Paglia, *Theor. Chim. Acta* **8**, 185 (1967).
Spectroscopic constants.

59. R.A. Hegstrom and W.N. Lipscomb, *J. chem. Phys.* **46**, 1594 (1967).
Magnetic properties.

60. J.R. de la Vega, Y. Fang, and H.F. Hameka, *Physica, 's Grav.* **36**, 577 (1967).
Magnetic properties.

61. J.M. O'Hare and R.P. Hurst, *J. chem. Phys.* **46**, 2356 (1967).
Hyperpolarizability.

62. V. Magnasco and A. Perico, *J. chem. Phys.* **47**, 971 (1967).
Localized orbitals.

63. J. Gruninger and H.F. Hameka, *Chem. Phys. Lett.* **1**, 14 (1967).
Magnetic properties.

64. P.E. Cade and W.M. Huo, *J. chem. Phys.* **47**, 614 (1967).
Hartree–Fock limit.

65. J.R. Hoyland, *J. chem. Phys.* **47**, 1556 (1967).
Ground state wave function.

66. A.A. Frost, B.H. Prentice, and R.A. Rouse, *J. Am. chem. Soc.* **89**, 3064 (1967).
Floating spherical gaussians.

67. J. Thorhallsson, *Z. Naturf.* **22a**, 1222 (1967).
V.B. and M.O.

68. R.F.W. Bader, I. Keaveny, and P.E. Cade, *J. chem. Phys.* **47**, 3381 (1967).
Charge distribution.

69. A.A. Frost, *J. chem. Phys.* **47**, 3707 (1967).
Floating spherical gaussians.

70. A.A. Frost, *J. chem. Phys.* **47**, 3714 (1967).
Bond length.

71. E.R. Davidson and C.F. Bender, *J. chem. Phys.* **49**, 465 (1968).
Contains discussion of pair correlation.

72. W.A. Sanders and M. Krauss, *J. Res. natn. Bur. Stand.* **72A**, 85 (1968).
Gaussian basis.

73. R. Ahlrichs and W. Kutzelnigg, *J. chem. Phys.* **48**, 1819 (1968).
Natural orbitals.

74. R.E. Brown and H. Shull, *Int. J. Quantum Chem.* **2**, 663 (1968).
Several excited states.

75. F. Grimaldi, A. Lecourt, and C.M. Moser, *Symp. Faraday Soc.* **2**, 59 (1968).
Dipole moment.

76. C.F. Bender and E.R. Davidson, *Phys. Rev.* **183**, 23 (1969).
Uses 939 configurations.

77. G.N. Avgeropoulos and D.D. Ebbing, *J. chem. Phys.* **50**, 3493 (1969).
Electric field gradient.

78. S.F. Boys and N.C. Handy, *Proc. R. Soc.* **A311**, 309 (1969).
Solution of molecular transcorrelated wave equation.

79. R.F.W. Bader and J.L. Ginsburg, *Can. J. Chem.* **47**, 3061 (1969).
Force constant.

80. J.A. Keefer, J.K. Su Fu, and R.L. Belford, *J. chem. Phys.* **50**, 160 (1969).
One-centre calculation.

81. W.E. Palke and W.A. Goddard, *J. chem. Phys.* **50**, 4524 (1969).
Generalized valence bond.

82. R.A. Rouse and A.A. Frost, *J. chem. Phys.* **50**, 1705 (1969).
Floating spherical gaussian.

83. J. Thorhallsson and D.P. Chong, *Chem. Phys. Lett.* **4**, 405 (1969).
Virial scaling of potential curve.

84. R.C. Sahni, B.C. Sawhney, and M.J. Hanley, *Trans. Faraday. Soc.* **65**, 3121 (1969).

85. J. Goodisman, *J. chem. Phys.* **51**, 3540 (1969).
Correlated wave function by perturbation theory.

LiH^+

State	Internuclear distance	Basis set	Energy	Wave function	Reference
$X\,^2\Sigma^+$	3·01	Valence bond STO	−7·699		1
	3·014	Ext. STO	−7·72943		3
	3·736	Elliptic orbitals valence bond	−7·758855		4
	4·25	STO + Elliptic fns. Valence bond	−7·7808	√	2

Property	*References*
Ionization potential	1, 2

REFERENCES

1. O. Platas, R.P. Hurst, and F.A. Matsen, *J. chem. Phys.* **31**, 501 (1959).
Valence bond calculation.

2. J.C. Browne, *J. chem. Phys.* **41**, 3495 (1964).
Valence bond.

3. P.E. Cade and W.M. Huo, *J. chem. Phys.* **47**, 614 (1967).
Hartree–Fock limit.

4. C.S. Liu, *J. chem. Phys.* **50**, 2787 (1969).
Very extensive V.B. calculation.

MgF

State	Internuclear distance	Basis set	Energy	Wave function	Reference
$X^2\Sigma^+$	3·194	Ext. STO			1
$A^2\Pi_r$	3·250	Ext. STO			1
$^2\Pi_i$		Ext. STO	repulsive		1

Properties	*References*
Potential curve	1
Spectroscopic constants	1
Excited states	1

REFERENCE

1. T.E.H. Walker and W.G. Richards, *J. Phys.* B**1**, 1061 (1968).
Discussion of nature of $A^2\Pi$ state.

MgH

State	Internuclear distance	Basis set	Energy	Wave function	Reference
$X^2\Sigma^+$	3·271	Ext. STO	−200·1566	✓	2

Properties	*References*
Dipole moment	1
Potential curve	2
Orbital energies	2
Spin–orbit coupling	3
Charge density	4

REFERENCES

1. P.E. Cade and W.M. Huo, *J. chem. Phys.* **45**, 1063 (1966).
Dipole moment.

2. P.E. Cade and W.M. Huo, *J. chem. Phys.* **47**, 649 (1967).
Near Hartree–Fock limit.

3. T.E.H. Walker and W.G. Richards, *Symp. Faraday Soc.* **2**, 64 (1968).
Spin–orbit coupling constant of $A\,^2\Pi$ state.

4. P.E. Cade, R.F.W. Bader, W.H. Henneker, and J. Keaveny *J. chem. Phys.* **50**, 5313 (1969).
Charge distribution with wave function of 2.

MgH^+

State	Internuclear distance	Basis set	Energy	Wave function	Reference
$X\,^1\Sigma^+$	3·271	Ext. STO	–199·9082		1

REFERENCE

1. P.E. Cade and W.M. Huo, *J. chem. Phys.* **47**, 649 (1967).
Near Hartree–Fock limit.

MgO

State	Internuclear distance	Basis set	Energy	Wave function	Reference
$^1\Sigma^+$	3·200	Ext. STO	–274·3754	✓	2
	3·3052	Ext. STO	–274·3862		3
	3·3052	Ext. STO	–274·3862	✓	4

Properties	*References*
Spectroscopic constants	1, 4
Potential curves	1, 2
Dipole moment	3, 4
Quadrupole coupling	4
Excited states	1

REFERENCES

1. W.G. Richards, G. Verhaegen, and C.M. Moser, *J. chem. Phys.* **45**, 3226 (1966).
Variational calculation to predict ground state.

2. A.D. McLean and M. Yoshimine, Supp. to *I.B.M. Jl Res. Dev.* (1967).

3. M. Yoshimine and A.D. McLean, *Int. J. Quantum Chem.* **1S**, 313 (1967).
Very near Hartree–Fock limit.

4. M. Yoshimine, *J. phys. Soc. Japan* **25**, 1100 (1968).
Very near Hartree–Fock limit.

$\underline{N_2}$

State	Internuclear distance	Basis set	Energy	Wave function	Reference
$X\,^1\Sigma_g^+$	2·0675	Min. STO	–108·5736	✓	21
	2·068	Min. STO	–108·63359	✓	5
	2·068	Min. STO	–108·63359		14
	2·095	STO	–108·63377	✓	35
	2·0866	Min. STO + CI	–108·63431		9
	2·068	Min. STO + CI	–108·66054		13
	2·068	STO	–108·74527	✓	12
	2·0675	Double Zeta	–108·7853	✓	10
	2·068	Ext. STO	–108·9714	✓	18
	2·026	Ext. STO	–108·97303		23
	2·0675	STO	–108·98488		24
	2·0134	Ext. STO	–108·9956	✓	25
$A\,^3\Sigma^+$	2·0675	Min. STO	–108·3091		21
	2·355	Min. STO	–108·45980		9
$^3\Delta_u$	2·359	Min. STO	–108·41606		9
$^5\Sigma_g^+$	2·750	Min. STO	–108·40874		9
$B\,^3\Pi_g$	2·0675	Min. STO	–108·2983		21
	2·289	Min. STO	–108·37526		9
$Y\,^3\Sigma_u^-$	2·361	Min. STO	–108·37233		9
$a'\,^1\Sigma^-$	2·361	Min. STO	–108·37233		9
$^5\Pi_u$	2·708	Min. STO	–108·34161		9
$w'\ \Delta_g$	2·355	Min. STO	–108·34057		9
$a'\ \Pi_g$	2·315	Min. STO	–108·30346		9
$C\,^3\Pi_u$	2·0675	Min. STO	–108·1086	✓	21
	2·138	Min. STO	–108·16430		9

Properties	*References*
Spectroscopic constants	1, 3, 9, 18, 22, 23, 25, 35
Quadrupole moment	2, 6, 18
Population analysis	2, 8, 13, 14, 22, 29
Orbital energies	2, 5, 11, 12, 14, 18, 21, 25, 28

Properties	*References*
Dissociation energies	2,5,10,30
Ionization potentials	2,21
Term values	2
Binding energy	3
Magnetic properties	4,7,15,17,19,27,33,37
Potential curve	9,11,18,20,23,25,35
Term values	9,10,21,23
Electric polarizability	16,34
Charge density maps	26,30,31
Orbital forces	30
Verdet constant	36

REFERENCES

1. H. Kopineck, *Z. Naturf.* **7a**, 22, 314 (1952).
V.B. treatment of $X\,^1\Sigma^+$ with STO basis.

2. C.W. Scherr, *J. chem. Phys.* **23**, 569 (1955).
Min. STO basis for $X\,^1\Sigma^+$. Wave functions given.

3. A.C. Hurley, *Proc. phys. Soc.* **A69** 767 (1956).
Wave functions as in ref. 2. CI included and intra-atomic electron correlation.

4. J. Baudet, J. Tillieu, and J. Guy, *C. r. hebd. Séanc. Acad. Sci., Paris*, **244**, 1756 (1957).
Wave functions as in ref. 2.

5. B.J. Ransil, *Rev. mod. Phys.* **32**, 239, 245 (1960).

6. J.W. Richardson, *Rev. mod. Phys.* **32**, 461 (1960).
Wave function as in refs. 5 and 10.

7. H.F. Hameka, *J. chem. Phys.* **34**, 366 (1961).
Wave functions as in ref. 2.

8. S. Fraga and B.J. Ransil, *J. chem. Phys.* **34**, 727 (1961).
Wave functions as in ref. 5.

9. S. Fraga and B.J. Ransil, *J. chem. Phys.* **35**, 669 (1961).
Limited CI included.

10. J.W. Richardson, *J. chem. Phys.* **35**, 1829 (1961).
Excited states $A\,^3\Sigma_u^+$, $^3\Delta_u$, $a'\,^1\Sigma_u^-$, $Y\,^3\Sigma_u^-$, $w'\,^1\Delta_u$, $b'\,^1\Sigma_u^+$.
Correlation energy estimated.

11. R.K. Nesbet, *Phys. Rev.* **122**, 1497 (1961).
STO + CI for $X\,^1\Sigma_g^+$.

12. E. Clementi, *Gazz. chim. ital.* **91**, 722 (1961).

13. S. Fraga and B.J. Ransil, *J. chem. Phys.* **36**, 1127 (1962).
Wave functions as in ref. 5.

14. E. Clementi and H. Clementi, *J. chem. Phys.* **36**, 2824 (1962).
Wave functions as in ref. 5.

15. M. Karplus and H.J. Kolker, *J. chem. Phys.* **38**, 1263 (1963).
Variation perturbation method using wave functions in ref. 5.

16. H.J. Kolker and M. Karplus, *J. chem. Phys.* **39**, 2011 (1963).
Variation perturbation method using wave functions in ref. 5.

17. H. Kim, H. Hamano, and H.F. Hameka, *Physica, 's Grav.* **29**, 117 (1963).
Wave functions as in ref. 2.

18. R.K. Nesbet, *J. chem. Phys.* **40**, 3619 (1964).

19. H.J. Kolker and M. Karplus, *J. chem. Phys.* **41**, 1259 (1964).
Variation perturbation method using wave functions in ref. 5.

20. L.M. Huber and W. Thorsen, *J. chem. Phys.* **41**, 1829 (1964).
V.B. Method using STO basis for $A\,^3\Sigma^+ - X\,^1\Sigma^+$ excitation energy.

21. R.C. Sahni and E.J. De Lorenzo, *J. chem. Phys.* **42**, 3612 (1965).
Spin polarization and restricted treatment

22. H. Lefebvre-Brion and C.M. Moser, *J. chem. Phys.* **43**, 1394 (1965).
Calculation of Rydberg levels with STO basis.

23. R.K. Nesbet, *J. chem. Phys.* **43**, 4403 (1965).
$^1\Pi_g$, $^1\Delta_u$, $^1\Sigma_u^-$, $^3\Sigma_u^-$, $^3\Delta_u$, $^3\Sigma_u^+$, and $^3\Pi_g$ treated by virtual orbital method.

24. F. Grimaldi, *J. chem. Phys.* **43**, S59 (1965).
Correlation energy calculation for $X\,^1\Sigma_g^+$ using CI.

25. P.E. Cade, K.D. Sales, and A.C. Wahl, *J. chem. Phys.* **44**, 1973 (1966).
Hartree–Fock limit.

26. A.C. Wahl, *Science, N.Y.* **151**, 961 (1966).
STO basis.

27. G.P. Arrighini, F. Grossi, and M. Maestro, *Theor. Chim Acta,* **5**, 266 (1966).
Perturbation treatment using wave functions in ref. 5.

28. R.J. Buenker, S.D. Peyerimhoff and J.L. Whitten, *J. chem. Phys.* **46**, 2029 (1967).
Comparison with C_2, O_2, and F_2 using correlation diagram. Wave functions given in ref. 25.

29. E.R. Davidson, *J. chem. Phys.* **46**, 3320 (1967).
Wave functions in refs. 5 and 25.

30. R.F.W. Bader, W.H. Henneker and P.E. Cade, *J. chem. Phys.* **46**, 3341 (1967).
Wave functions as in ref. 25.

31. B.J. Ransil and J.J. Sinai, *J. chem. Phys.* **46**, 4050 (1967).
STO basis.

32. V. Magnasco and A. Perico, *J. chem. Phys.* **47**, 971 (1967).
Localization of molecular and atomic orbitals using wave functions in ref. 5.

33. J.R. de la Vega, *J. chem. Phys.* **47**, 1834 (1967).
Wave functions in ref. 5.

34. J.M. O'Hare and R.P. Hurst, *J. chem. Phys.* **46**, 3356 (1967).
Wave functions in refs. 5 and 18.

35. R.C. Sahni and B.C. Sawkney, *Int. J. Quantum Chem.* **1**, 257 (1967).

36. Y. I'Haya, *Int. J. Quantum Chem.* **1**, 693 (1967).
Wave functions as in ref. 10.

37. J.R. de la Vega and H.F. Hameka, *Physica, 's Grav.* **35**, 313 (1967).
Wave functions as in ref. 5.

38. M.A. Marchetti and S.R. La Paglia, *J. chem. Phys.* **48**, 434 (1968).
$^1\Sigma_u^+ - {}^1\Sigma_g^+$ dipole strengths using wave functions in ref. 5 and CI.

39. W.H. Henneker and P.E. Cade, *Chem. Phys. Lett.* **2**, 575 (1968).
Electron distribution in momentum space.

N_2^+

State	Internuclear distance	Basis set	Energy	Wave function	Reference
$X\,^2\Sigma_g^+$	2·0675	Min. STO	−108·0294	√	3
	2·177	Min. STO	−108·08962		1
	2·111	STO	−108·1260	√	9
	2·075	Ext. STO	−108·33474		5
	2·0385	Ext. STO	−108·4079	√	6
$A\,^2\Pi_u$	2·231	STO	−108·12100	√	9
	2·249	Min. STO	−108·12259		1
	2·0675	Double zeta	−108·2067	√	2
	2·177	Ext. STO	−108·36337		5
	2·134	Ext. STO	−108·432	√	6
$B\,^2\Sigma_u^+$	2·0675	Min. STO	−107·8432	√	3
	1·992	STO	−107·95030	√	9
	1·935	Ext. STO	−108·2702	√	6
$^2\Pi_g$	2·482	Min. STO	−107·76767		1

Properties	*References*
Spectroscopic constants	1, 4, 5, 6, 9
Term values	1, 2, 3, 5, 8
Potential curve	1, 5, 6, 9

Properties	*References*
Ionization potentials	3
Orbital energies	3, 6
Orbital forces	7

REFERENCES

1. S. Fraga and B.J. Ransil, *J. chem. Phys.* **35**, 669 (1961). Predicts $^2\Pi_u$ ground state.
2. J.W. Richardson, *J. chem. Phys.* **35**, 1829 (1961). Predicts $^2\Pi_u$ ground state.
3. R.C. Sahni and E.J. De Lorenzo, *J. chem. Phys.* **42**, 3612 (1965). Spin polarization and restricted treatment.
4. H. Lefebure-Brion and C.M. Moser, *J. chem. Phys.* **43**, 1394 (1965). Min. STO basis for $X\,^2\Sigma_g^+$, $A\,^2\Pi_u$, and $B\,^2\Sigma_u^+$.
5. R.K. Nesbet, *J. chem. Phys.* **43**, 4403 (1965). Virtual orbital method. Predicts $^2\Pi_u$ ground state.
6. P.E. Cade, K.D. Sales, and A.C. Wahl, *J. chem. Phys.* **44**, 1973 (1966). Predicts $^2\Pi_u$ ground state. Hartree–Fock limit.
7. R.F.W. Bader, W.H. Henneker and P.E. Cade, *J. chem. Phys.* **46**, 3341 (1967). $X\,^2\Sigma_g^+$, $A\,^2\Pi_u$, $B\,^2\Sigma_u^+$ states. Wave functions as in ref. 6.
8. G. Verhaegen, W.G. Richards, and C.M. Moser, *J. chem. Phys.* **47**, 2595 (1967). $X\,^2\Sigma^+$, $A\,^2\Pi_u$ states from basis in R.K. Nesbet *J. chem. Phys.* **40**, 3619 (1964). Correlation energy estimated to correct reversal of states in other papers.
9. R.C. Sahni and B.C. Sawhney, *Int. J. Quantum Chem.* **1**, 251 (1967).

N_2^{++}

State	Internuclear distance	Basis set	Energy	Wave function	Reference
$X\,^1\Sigma_g^+$	2·338	Min. STO	–106·95030		1
	2·0675	Double zeta	–107·1019	✓	2

Properties	*References*
Term values	1
Spectroscopic constants	1

1. S. Fraga and B.J. Ransil, *J. chem. Phys.* **35**, 669 (1961).

2. J.W. Richardson, *J. chem. Phys.* **35**, 1829 (1961).

NF

State	Internuclear distance	Basis set	Energy	Wave function	Reference
$X^3\Sigma^-$	2·45	Min. STO	−153·2046		1
$b\ ^1\Sigma^+$	2·45	Min. STO	−153·1138		1

Properties	*Reference*
Potential curve	1
Excited states	1

REFERENCE

1. R.C. Sahni, *Trans. Faraday Soc.* **63**, 801 (1967).
Min. STO.

NF^+

State	Internuclear distance	Basis set	Energy	Wave function	Reference
$X\ ^2\Pi$	2·30	Min. STO	−152·8214		1
$^4\Pi$	2·90	Min. STO	−152·8248		1

Properties	*Reference*
Potential curves	1
Excited states	1

REFERENCE

1. R.C. Sahni, *Trans. Faraday Soc.* **63**, 801 (1967).
Min. STO.

NH

State	Internuclear distance	Basis set	Energy	Wave function	Reference
$X^3\Sigma^-$	2·0	Contracted GTO + CI	−54·5409		11

State	Internuclear distance	Basis set	Energy	Wave function	Reference
		Minimum basis One centre	–54·55581		13
	1·97	Min. STO	–54·7827	√	4
	1·90	One centre STO	–54·9064	√	12
	1·9729	Ext. STO	–54·97838	√	16
	1·9614	Ext. STO + CI (3379 configs)	–55·1620	√	22
$d\ ^1\Sigma^+$	1·976	Min. STO	–54·34506	√	7, 4

Properties	*References*
Dipole moment	2, 3, 4, 5, 7, 11, 12, 22
Spectroscopic constants	2, 3, 11, 16, 20
Charge distribution	2, 17
Excitation energies	2, 3, 4, 21
Magnetic properties	8, 14, 22
Spin–orbit interaction	15
Ionization potential	16
Electron affinity	18

REFERENCES

1. J. Higuchi, *J. chem. Phys.* **24**, 535 (1956).
Simplified calculation.

2. M. Krauss, *J. chem. Phys.* **28**, 1021 (1958).
Min. basis calc. on ground state.

3. M. Krauss and J. Wehner, *J. chem. Phys.* **29**, 1287 (1958).
CI on wave function of ref. 2.

4. M.E. Boyd, *J. chem. Phys.* **29**, 108 (1958).
Several excited states.

5. A.C. Hurley, *Proc. R. Soc.* **A248**, 119 (1958).
CI on Krauss wave function.

6. A.C. Hurley, *Proc. R. Soc.* **A249**, 402 (1959.
Semi-empirical account of excited states.

7. S. Fraga and B.J. Ransil, *J. chem. Phys.* **36**, 1127 (1962).
Min. basis + CI.

8. C.W. Kern and W.N. Lipscomb, *J. chem. Phys.* **37**, 260 (1962).
Magnetic shielding.

9. D.M. Bishop and J.R. Hoyland, *Molec. Phys.* **7**, 161 (1963).
Single-configuration valence bond.

10. C.M. Reeves, *J. chem. Phys.* **39**, 1 (1963).
Preliminary investigations.

11. C.M. Reeves and R. Fletcher, *J. chem. Phys.* **42**, 4073 (1965).
Variety of gaussian bases.

12. B.D. Joshi, *J. chem. Phys.* **43**, S40 (1965).
Extended basis single-centre calculation.

13. J.B. Lounsbury, *J. chem. Phys.* **42**, 1549 (1965).
Min. basis single centre.

14. J.B. Lounsbury, *J. chem. Phys.* **46**, 2193 (1967); *J. chem. Phys.* **47**, 1566 (1967).
Zero-field splitting.

15. J.W. McIver and H.F. Hameka, *J. chem. Phys.* **45**, 767 (1966).
+ erratum *J. chem. Phys.* **46**, 825 (1967). Spin–orbit interaction

16. P.E. Cade and W.M. Huo, *J. chem. Phys.* **47**, 614 (1967).
Hartree–Fock limit.

17. R.F.W. Bader, I. Keaveny, and P.E. Cade, *J. chem. Phys.* **47**, 3381 (1967).
Bonding characteristics.

18. P.E. Cade, *Proc. phys. Soc.* **91**, 842 (1967).
Electron affinity.

19. V. Magnasco and A. Perico, *J. chem. Phys.* **47**, 971 (1967).
Localized orbitals.

20. S.R. La Paglia, *Theor. Chim. Acta*, **8**, 185 (1967).
Transition probabilities.

21. P.E. Cade, *Can. J. Phys.* **46**, 1989 (1968).
Singlet triplet intercombination separation.

22. C.F. Bender and E.R. Davidson, *Phys. Rev.* **183**, 23 (1969).
Near experimental energy.

NH^+

State	Internuclear distance	Basis set	Energy	Wave function	Reference
$X\,^2\Pi_r$	2·048	Ext. STO	–54·50701		1

REFERENCE

1. P.E. Cade and W.M. Huo, *J. chem. Phys.* **47**, 614 (1967).

NH^-

REFERENCE

1. T.E.H. Walker and W.G. Richards, *Symp. Faraday Soc.* **2**, 68 (1968). Spin–orbit coupling constant.

NH^{--}

State	Internuclear distance	Basis set	Energy	Wave function	Reference
$^1\Sigma^+$	1·962	GTO	–54·5076		1

Property	*Reference*
Proton affinity	1

REFERENCE

1. A.C. Hopkinson, N.K. Holbrook, K. Yates, and I.G. Csizmadia, *J. chem. Phys.* **49**, 3596 (1968. Various basis sets.

NO

State	Internuclear distance	Basis set	Energy	Wave function	Reference
$X\,^2\Pi$	2·173	Min. STO + CI	–128·861	✓	1, 1a
		Min. STO + CI	–128·869		4

Property	*References*
Spectroscopic term values	5, 6, 8
Ionization potential	1, 5
Atomic populations	1, 5, 8
Spin–orbit coupling	2, 3, 5, 7, 8
Magnetic constants	2, 3, 4,
Quadrupole coupling	2, 3, 4, 5
Dipole moment	1, 5
Oscillator strengths	5
Transition probability	9

REFERENCES

1. H. Brion, C.M. Moser, and M. Yamazaki, *J. chem. Phys.* **30**, 673 (1959).

1a. H. Brion, C.M. Moser, and M. Yamazaki, *J. chem. Phys.* **33**, 1871 (1960).
Erratum to 1.

2. H. Lefebvre-Brion and C.M. Moser, *Phys. Rev.* **48**, 675 (1960).
Uses wave function of 1.

3. C.C. Lin, K. Hijikata, and M. Sakamoto, *J. chem. Phys.* **33**, 878 (1960).
Uses wave function of 1.

4. M. Yamazaki, M. Sakamoto, K. Hijikata, and C.C. Lin, *J. chem. Phys.* **34**, 1926 (1961).
Extensive CI on wave function of 1.

5. M. Yamazaki, *Sci. Rep. Kanazawa Univ.* **8**, 371 (1963).
Very extensive CI on wave function of 1 and excited states.

6. H. Lefebvre-Brion and C.M. Moser, *J. molec. Spectrosc.* **15**, 211 (1965).
Rydberg levels with virtual orbitals.

7. A. Hellman and C. Ballhausen, *Theor. Chim. Acta,* **3**, 159 (1965).
Spin–orbit coupling constant.

8. H. Lefebvre-Brion and C.M. Moser, *J. chem. Phys.* **44**, 2951 (1966).
Excitation energies and spin–orbit coupling.

9. S.H. Liu, *Theor. Chim. Acta,* **8**, 1 (1967).
Transition probability.

Na_2

Properties	*References*
Spectroscopic constants	1, 2, 3
Binding energy	3
Oscillator strengths	3
Excitation energy	3

REFERENCES

1. N. Rosen, *Phys. Rev.* **38**, 255 (1931).
Interaction energy of 2 normal Na atoms using STO basis (for outer electrons only) in V.B. approximation giving $X\ ^1\Sigma_g^+$ state.

2. N. Rosen and S. Ikehara, *Phys. Rev.* **43**, 5 (1933).
Interaction energy of 2 normal Na atoms using STO basis (for outer electrons only) in V.B. approximation giving $X\ ^1\Sigma_g^+$ state.

3. D.W. Davies, *Trans. Faraday. Soc.* **54**, 1429 (1958).
$^1\Sigma_g^+$, $^1\Sigma_u^+$ states considered using STO basis (outer electrons only) in V.B. approximation.

NaBr

State	Internuclear distance	Basis set	Energy	Wave function	Reference
$^1\Sigma^+$	4·728	Ext. STO	−2734·2876	✓	1

Properties	*Reference*
Orbital energies	1
Potential curve	1

REFERENCE

1. A.D. McLean and M. Yoshimine, Supp. to *I.B.M. Jl Res. Dev.* (1967).
Near Hartree–Fock limit.

NaCl

State	Internuclear distance	Basis set	Energy	Wave function	Reference
$^1\Sigma^+$	4·4609	Ext. STO	−621·4574	✓	1
	4·485	Ext. STO	−621·4974	✓	2

Properties	*References*
Orbital energies	1, 2
Potential curve	1, 2
Dipole moment	2
Quadrupole coupling	2
Spectroscopic constants	2

REFERENCES

1. A.D. McLean and M. Yoshimine, Supp. to *I.B.M. Jl. Res. Dev.* (1967).
Very near Hartree–Fock limit.

2. R.L. Matcha, *J. chem. Phys.* **48**, 335 (1968).
Very near Hartree–Fock limit. Many one-electron properties.

NaF

State	Internuclear distance	Basis set	Energy	Wave function	Reference
$^1\Sigma^+$	3·62883	Ext. STO	−261·3784	✓	1
	3·628	Ext. STO	−261·3785	✓	2

Properties	*References*
Orbital energies	1, 2
Potential curve	1, 2
Dipole moment	2
Quadrupole coupling	2
Spectroscopic constants	2

REFERENCES

1. A.D. McLean and M. Yoshimine, Supp. to *I.B.M. Jl Res. Dev.* (1967). Near Hartree–Fock limit.

2. R.L. Matcha, *J. chem. Phys.* **47**, 5295 (1967). Very near Hartree–Fock limit. Many one-electron properties.

3. H. Preuss and R. Janoschek, *J. molec. Struct.* **3**, 423 (1969). Preliminary calculation $E = -258{\cdot}7061$.

NaH

State	Internuclear distance	Basis set	Energy	Wave function	Reference
$^1\Sigma^+$	3·566	Ext. STO	−162·3928	✓	2

Properties	*References*
Dipole moment	1, 3
Potential curve	2
Spectroscopic constants	2
Orbital energies	2
Ionization potential	2
Force constant	4
Charge distribution	5

REFERENCES

1. P.E. Cade and W.M. Huo, *J. chem. Phys.* **45**, 1063 (1966). Dipole moment.

2. P.E. Cade and W.M. Huo, *J. chem. Phys.* **47**, 649 (1967).
Near Hartree–Fock limit.

3. F. Grimaldi, A. Lecourt, and C.M. Moser, *Symp. Faraday Soc.* **2**, 59 (1968).
Dipole moment with CI.

4. R.F.W. Bader and J.L. Ginsberg, *Can. J. Chem.* **47**, 3001 (1969).
Force constant.

5. P.E. Cade, R.F.W. Bader, W.H. Henneker, and I. Keaveny, *J. chem. Phys.* **50**, 5313 (1969).
Charge distribution.

NaH^+

State	Internuclear distance	Basis set	Energy	Wave function	Reference
$^2\Sigma^+$	3·566	Ext. STO	–162·1671		1

REFERENCE

1. P.E. Cade and W.M. Huo, *J. chem. Phys.* **47**, 649 (1967).
Near Hartree–Fock limit.

Ne_2

State	Internuclear distance	Basis set	Energy	Wave function	Reference
$X\ ^1\Sigma_g^+$	3·0	STO	–256·9957	✓	1

Properties	*Reference*
Potential curves	1
Orbital energies	1

REFERENCE

1. T.L. Gilbert and A.C. Wahl, *J. chem. Phys.* **47**, 3425 (1967).

NeF

Property	*Reference*
Potential curve	1

REFERENCE

1. L.C. Allen, A.M. Lesk, and R.M. Erdahl, *J. Am. chem. Soc.* **88**, 615 (1966).
V.B. wave functions for $X\,^1\Sigma^+$ given.

NeH^+

State	Internuclear distance	Basis set	Energy	Wave function	Reference
$^1\Sigma^+$	2·33	One centre	–127·5703		2
	1·90	Ext. GTO	–128·6029	✓	3
	1·83	Ext. STO	–128·6284		1

Properties	*References*
Potential curve	1, 3
Spectroscopic constants	1, 2
Orbital energies	3
Charge density	1

REFERENCES

1. S.D. Peyerimhoff, *J. chem. Phys.* **43**, 998 (1965).
Ext. STO.

2. K.E. Banyard and A. Sutton, *J. chem. Phys* **46**, 2143 (1967).
One centre with minimum basis set.

3. J.B. Moffat, *Can. J. Chem.* **46**, 3893 (1968).
GTO.

NeO

State	Internuclear distance	Basis set	Energy	Wave function	Reference
$^1\Sigma^+$		Valence bond	repulsive curve		1

Property	*Reference*
Potential curve	1

REFERENCE

1. L.C. Allen, A.M. Lesk, and R.M. Erdahl, *J. Am. chem. Soc.* **88**, 615 (1966).

$\underline{O_2}$

State	Internuclear distance	Basis set O_2	Energy	Wave function	Reference
$X^3\Sigma_g^-$	2·3	STO + CI	−149·00734	√	2
	2·282	Min. STO	−149·0921	√	4
	2·5	Min. STO	−149·2205		12
$a\ ^1\Delta_g$	2·3	STO + CI	−148·9528		2
	2·282	Min. STO	−149·0105	√	4
$^1\Sigma_g^+$	2·3	STO + CI	−148·8982		2
$^1\Sigma_u^-$	2·3	STO + CI	−148·8909		2
$^3\Sigma_u^+$	2·3	STO + CI	−148·8816		2
$^3\Sigma_u^-$	2·3	STO + CI	−148·5228		2

Properties	*References*
Spectroscopic constants	1, 2, 11, 12, 14
Potential curve	1, 11, 12
Population analysis	2
Quadrupole moment	2
Magnetic properties	2
Polarizabilities	2
Orbital energies	2, 4, 8
Term values	2, 4, 11, 12
Spin–spin splitting	3, 5, 9
Ionization potential	4
Charge density maps	6, 10
Dissociation energy	10
Orbital forces	10

REFERENCES

1. A. Meckler, *J. chem. Phys.* **21**, 1750 (1953).
   CI on GTO basis for $X\ ^3\Sigma_g^-$, $^1\Sigma_g^+$. Wave functions given.

2. M. Kotani, Y. Mizumo, K, Kayama, and E. Ishiguro, *J. phys. Soc. Japan*, **12**, 707 (1957).
   Several methods of calculation used.

3. K. Kayama, *J. chem. Phys.* **42**, 622 (1965).
   Wave functions as in ref. 2.

4. R.C. Sahni and E.J. De Lorenzo, *J. chem. Phys.* **42**, 3612 (1965).
   Spin polarization and restricted treatment.

5. K. Kayama and J.C. Baird, **43**, 1082 (1965).
   Approximations in ref. 3 used.

6. A.C. Wahl, *Science, N.Y.* **151**, 961 (1966).
   STO basis for $X\ ^3\Sigma_g^-$.

7. M. Bishop and J. Arenti, *J. Phys. Chem.* **70**, 3748 (1966).
Perturbation treatment using wave functions in ref. 2 to give energy difference between $X\,^3\Sigma_g^-$ and $a\,^1\Delta_g$.

8. R.J. Buenker, S.D. Peyerimhoff, and J.L. Whitten, *J. chem. Phys.* **46**, 2029 (1967).
Comparison with C_2, N_2, and F_2 using correlation diagram. Wave functions due to P.E. Cade and G. Malli (unpub.)

9. K. Kayama and J.C. Baird, *J. chem. Phys.* **46**, 2604 (1967).
Wave functions as in refs. 1 and 2.

10. R.F.W. Bader, W.H. Henneker, P.E. Cade, *J. chem. Phys.* **46**, 3341 (1967).
Wave functions due to P.E. Cade and G. Malli (unpub.).

11. H.F. Schaeffer and P.E. Harris, *Chem. Phys. Lett.* **1**, 407 (1967).
CI using open shell spin projected functions (UCI). Min. STO for $X\,^3\Sigma_g^-$, $a\,^1\Delta_g$, and $^1\Sigma_g^+$.

12. H.F. Schaeffer and F.E. Harris, *J. chem. Phys.* **48**, 4946 (1968).
Calculation on 62 low-lying states of O_2 resulting from 'O' atoms in ^3P, ^1D, ^1S states.

13. H. Taketa, H. Takewaki, O. Namura and K. Ohno, *Theor. Chim. Acta*, **11**, 369 (1969).
Preliminary results for unrestricted Hartree–Fock treatment of $^3\Sigma_\omega^- - {}^3\Sigma_\mu^+$ transition.

14. J. Leclerc, *Annls Astrophys.* **30**, 93 (1967).
Rydberg levels.

O_2^+

State	Internuclear distance	Basis set	Energy	Wave function	Reference
$X\,^2\Pi_g$	2·28167	Min. STO	−148·6941	✓	2
$a\,^4\Pi_u$	2·28167	Min. STO	−148·6682	✓	2

Properties	*References*
Term values	1, 2
Ionization potentials	2
Orbital energies	2
Orbital forces	3

REFERENCES

1. M. Kotani, Y. Mizumo, K. Kayama, and E. Ishiguro, *J. phys. Soc. Japan*, **12**, 707 (1957).
STO + CI for $^4\Sigma_g^-$, $^4\Pi_u$, $^2\Pi_g$, and $^2\Pi_u$ states.

2. R.C. Sahni and E.J. DeLorenzo, *J. chem. Phys.* **42**, 3612 (1965).
Spin polarization and restricted treatment.

3. R.F.W. Bader, W.H. Henneker, and P.E. Cade, *J. chem. Phys.* **46**, 3341 (1967).
P.E. Cade and G. Malli wave functions for $X\,^2\Pi_g$ (unpub.).

4. F. Bassani, E. Montaldi, and F.G. Fumi, *Nuovo Cim.* **3**, 893 (1956).

O_2^-

Property	*Reference*
Orbital forces	1

REFERENCE

1. R.F.W. Bader, W.H. Henneker, and P.E. Cade, *J. chem. Phys.* **46**, 3341 (1967).
P.E. Cade and G. Malli wave functions for $^2\Pi_g$ (unpub.).

OH

State	Internuclear distance	Basis set	Energy	Wave function	Reference
$X\,^2\Pi_i$	1·8342	Min. STO Valence bond	−75·062	✓	2
	1·8342	+ CI	−75·327	✓	3
	1·795	Ext. STO	−75·42127	✓	12
	1·8342	Ext. STO + CI (2401 configs.)	−75·6422		16

Properties	*References*
Dipole moment	1, 2, 3
Spectroscopic properties	1
Charge distribution	1, 3, 13
Excitation energies	1, 6
Hyperfine structure	8, 11, 17
Spin–orbit coupling	14, 15

REFERENCES

1. M. Krauss, *J. chem. Phys.* **28**, 1021 (1958).
Min. basis set.

2. M. Krauss and J.F. Wehner, *J. chem. Phys.* **29**, 1287 (1958).
CI on Krauss wave function.

3. A.J. Freeman, *J. chem. Phys.* **28**, 230 (1958).
Hartree–Fock orbitals as basis.

4. R. Gáspár and I. Tamássy-Lentei, *Acta phys. hung.* **9**, 105 (1958).
Single centre.

5. A.C. Hurley, *Proc. R. Soc.* **A248**, 119 (1958).
CI on Krauss function.

6. A.C. Hurley, *Proc. R. Soc.* **A249**, 402 (1959).
Semi-empirical treatment of excited states.

7. A.J. Freeman, *Rev. mod. Phys.* **32**, 273 (1960).
Compares open-shell methods.

8. K. Kayama, *J. chem. Phys.* **39**, 1507 (1963).
Hyperfine structure constants.

9. D.M. Bishop and J.R. Hoyland, *Molec. Phys.* **7**, 161 (1963).
Single-centre calculation.

10. K. Pecul, *Acta phys. pol.* **27**, 713 (1965).
Min. basis V.B.

11. A.L.H. Chung, *J. chem. Phys.* **46**, 3144 (1966).
Isotopic hyperfine splitting constant.

12. P.E. Cade and W.M. Huo, *J. chem. Phys.* **47**, 614 (1967).
Hartree–Fock limit.

13. R.F.W. Bader, I. Keaveny, and P.E. Cade, *J. chem. Phys.* **47**, 3381 (1967).

14. T.E.H. Walker and W.G. Richards, *Symp. Faraday Soc.* **2**, 64 (1968).

15. T.E.H. Walker and W.G. Richards, *Phys. Rev.* **177**, 100 (1969).
Spin–orbit coupling.

16. C.F. Bender and E.R. Davidson, *Phys. Rev.* **183**, 23 (1969).
Near experimental energy.

17. W. Meyer, *J. chem. Phys.* **51**, 5149 (1969).

OH^-

State	Internuclear distance	Basis set	Energy	Wave function	Reference
$X\,^1\Sigma^+$	1·8103	Valence bond STO	−74·802		4
	1·8103	Min. STO	−74·8221	✓	5
	1·8342	Min. STO	−74·8785	✓	2
	1·814	Ext. GTO	−75·30		14

State	Internuclear distance	Basis set	Energy	Wave function	Reference
	1·82	Contracted GTO	−75·305		9
	1·80	Ext. GTO	−75·3670		7
	1·809	Ext. GTO	−75·3756		11
	1·79	Ext. GTO	−75·3961		8
	1·781	Ext. STO	−75·41754	✓	10

Properties	*References*
Dipole moment	10
Quadrupole moment	10
Charge density	10
Spectroscopic properties	9, 10, 14
Electron affinity	10
Proton affinity	11
Excitation energy	13

REFERENCES

1. R. Gáspár and I. Tamássy-Lentei, *Acta phys. hung.* **9**, 105 (1958).
United atom.

2. M. Krauss and B.J. Ransil, *J. chem. Phys.* **33**, 840 (1960).
Minimum basis set calculation.

3. R. Gáspár, I. Tamássy-Lentei, and Y. Kruglyak, *J. chem. Phys.* **36**, 740 (1962).
One-centre calculation.

4. R. Grahn, *Ark. Fys.* **28**, 85 (1965).
Single config. V.B. calculation.

5. J.L.J. Rosenfeld, *J. chem. Phys.* **40**, 384 (1964).
Minimum basis.

6. K. Pecul, *Acta phys. pol.* **27**, 713 (1965).
Single configuration V.B. one centre.

7. J.W. Moskowitz and M.C. Harrison, *J. chem. Phys.* **43**, 3550 (1965).
Extended Gaussian basis.

8. C.D. Ritchie and H.F. King, *J. chem. Phys.* **47**, 564 (1967).

9. R. Janoschek, H. Preuss, and G. Diercksen, *Int. J. Quantum Chem.* **1**, 649 (1967).

10. P.E. Cade, *J. chem. Phys.* **47**, 2390 (1967).

11. A.C. Hopkinson, N.K. Holbrook, K. Yates, and I.G. Csizmadia, *J. chem. Phys.* **49**, 3596 (1968).
Various basis sets.

12. A.A. Frost, *J. phys. Chem., Ithaca,* **72**, 1289 (1968).
Floating spherical gaussians.

13. P.E. Cade, *Can. J. Phys.* **46**, 1989 (1968).
Singlet triplet intercombination.

14. H. Preuss and R. Janoschek, *J. molec. Struct.* **3**, 423 (1969).

P_2

State	Internuclear distance	Basis set	Energy	Wave function	Reference
$X\,^1\Sigma_g^+$	3·580	Min. STO	679·1664	✓	1

Properties	*Reference*
Orbital energies	1
Population analysis	1

REFERENCE

1. D.B. Boyd and W.N. Lipscomb, *J. chem. Phys.* **46**, 910 (1967).

PH

State	Internuclear distance	Basis set	Energy	Wave function	Reference
$^3\Sigma^-$	2·708	Ext. STO	–341·2932	✓	2

Properties	*References*
Dipole moment	1
Orbital energies	2
Spectroscopic constants	2
Potential curve	2
Ionization potentials	2
Singlet–triplet separations	3
Charge distribution	4

REFERENCES

1. P.E. Cade and W.M. Huo, *J. chem. Phys.* **45**, 1063 (1966).

2. P.E. Cade and W.M. Huo, *J. chem. Phys.* **47**, 649 (1967).
Near Hartree–Fock limit.

3. P.E. Cade, *Can. J. Phys.* **46**, 1989 (1968).
Singlet–triplet separation.

4. P.E. Cade, R.F.W. Bader, W.H. Henneker, and I. Keaveny, *J. chem. Phys.* **50**, 5313 (1969).
Charge distribution.

PH^+

State	Internuclear distance	Basis set	Energy	Wave function	Reference
$X\,^2\Pi$	2·708	Ext. STO	−340·9384		1

Property	*References*
Electron affinity	1

REFERENCE

1. P.E. Cade and W.M. Huo, *J. chem. Phys.* **47**, 649 (1967).
Near Hartree–Fock limit.

PH^-

State	Internuclear distance	Basis set	Energy	Wave function	Reference
$^2\Pi$	2·668	Ext. STO	−341·2849	✓	1

Properties	*References*
Orbital energies	1
Electron affinity of PH	1
Spin–orbit coupling constant	2

REFERENCES

1. P.E. Cade, *Proc. phys. Soc.* **91**, 842 (1967).
Near Hartree–Fock limit.

2. T.E.H. Walker and W.G. Richards, *Symp. Faraday Soc.* **2**, 64 (1968).
Calculates spin–orbit coupling constant.

PN

State	Internuclear distance	Basis set	Energy	Wave function	Reference
$X\,^1\Sigma^+$	2·818	Ext. STO	−395·1848		1
	2·67	Ext. STO	−395·1857	√	2

Properties	*References*
Potential curve	2
Orbital energies	2
Dipole moment	1

REFERENCES

1. M. Yoshimine and A.D. McLean, *Int. J. Quantum Chem.* **1S**, 313 (1967).
2. A.D. McLean and M. Yoshimine, Supp. to *I.B.M. Jl Res. Dev.* (1967).
   Near Hartree–Fock limit.

PO

State	Internuclear distance	Basis set	Energy	Wave function	Reference
$X\,^2\Pi$	2·738	Min. STO	−414·1371	√	1

Properties	*References*
Dipole moment	1
Orbital energies	1
Population analysis	1

REFERENCE

1. D.B. Boyd and W.N. Lipscomb, *J. chem. Phys.* **46**, 910 (1967).

PO^-

State	Internuclear distance	Basis set	Energy	Wave function	Reference
$^3\Sigma^-$	2·859	Min. STO	−414·1168	√	1

Properties	*References*
Dipole moment	1
Orbital energies	1
Population analysis	1

REFERENCE

1. D.B. Boyd and W.N. Lipscomb, *J. chem. Phys.* **46**, 910 (1967).

RbF

State	Internuclear distance	Basis set	Energy	Wave function	Reference
$X\,^1\Sigma^+$	4·363	Ext. STO	3937·7727	✓	1

Property	*Reference*
Orbital energies	1

REFERENCE

1. A.D. McLean and M. Yoshimine, Supp. to *I.B.M. Jl Res. Dev.* (1967). Approaches Hartree–Fock limit.

SH

State	Internuclear distance	Basis set	Energy	Wave function	Reference
$X\,^2\Pi_i$	2·551	Ext. STO	398·1015	✓	2

Properties	*References*
Dipole moment	1
Orbital energies	2
Spectroscopic constants	2
Ionization potentials	2
Spin–orbit coupling constant	3
Charge distribution	4

REFERENCES

1. P.E. Cade and W.M. Huo, *J. chem. Phys.* **45**, 1063 (1966).

2. P.E. Cade and W.M. Huo, *J. chem. Phys.* **47**, 649 (1967). Near Hartree–Fock limit.

3. T.E.H. Walker and W.G. Richards, *Symp. Faraday Soc.* **2**, 64 (1968).
Calculation of spin–orbit coupling constant.

4. P.E. Cade, R.F.W. Bader, W.H. Henneker, and I. Keaveny, *J. chem. Phys.* **50**, 5313 (1969).
Charge distribution.

SH^+

State	Internuclear distance	Basis set	Energy	Wave function	Reference
$X\,^3\Sigma^-$	2·551	Ext. STO	−397·7593		1

Property	*Reference*
Singlet–triplet separation	2

REFERENCES

1. P.E. Cade and W.M. Huo, *J. chem. Phys.* **47**, 649 (1967).
Near Hartree–Fock limit.

2. P.E. Cade, *Can. J. Phys.* **46**, 1989 (1968).
Singlet–triplet separation using wave function of 1.

SH^-

State	Internuclear distance	Basis set	Energy	Wave function	Reference
$X\,^1\Sigma^+$	2·523	Min. GTO	−379·9070		3
	2·464	Min. STO One centre	−394·2616		1
	2·551	Ext. STO	−398·1459	✓	2

Properties	*References*
Orbital energies	2
Spectroscopic constants	2
Proton affinity	3

REFERENCES

1. K.E. Banyard and R.B. Hake, *J. chem. Phys.* **41**, 3221 (1964).
One-centre.

2. P.E. Cade, *J. chem. Phys.* **47**, 2390 (1967).
Near Hartree–Fock limit.

3. A.C. Hopkinson, N.K. Holbrook, K. Yates, and I.G. Csizmadia, *J. chem. Phys.* **49**, 3596 (1968).
Proton affinity calculated for S^{--}

ScF

State	Internuclear distance	Basis set	Energy	Wave function	Reference
$^1\Sigma^+$	3·31	Small STO	−858·54536	√	1
$^3\Delta$	3·31	Small STO	−858·56373		1

Properties	*References*
Orbital energies	1
Spectroscopic constants	1
Excited states	1
Charge density	2

REFERENCES

1. K.D. Carlson and C.M. Moser, *J. chem. Phys.* **46**, 35 (1967).
After CI $^1\Sigma^+$ is lower than $^3\Delta$.

2. I. Cohen and K.D. Carlson, *J. phys. Chem., Ithaca*, **73**, 1356 (1969).
Charge density.

ScO

State	Internuclear distance	Basis set	Energy	Wave function	Reference
$X\,^2\Sigma^+$	3·05	STO	−888·178	√	1

Properties	*References*
Spectroscopic constants	1
Potential curve	1
Excited states	1
Dipole moment	1

REFERENCE

1. K.D. Carlson, E. Ludena, and C. Moser, *J. chem. Phys.* **43**, 2408 (1965).
Discussion of excited states.

SiH

State	Internuclear distance	Basis set	Energy	Wave function	Reference
$X\,^2\Pi_r$	2·874	Ext. STO	−289·4362	✓	2

Properties	*References*
Dipole moment	1
Orbital energies	2
Spectroscopic constants	2
Potential curves	2
Ionization potentials	2
Spin–orbit coupling constant	3
Charge distribution	4

REFERENCE

1. P.E. Cade and W.M. Huo, *J. chem. Phys.* **45**, 1063 (1966).
2. P.E. Cade and W.M. Huo, *J. chem. Phys.* **47**, 649 (1967). Near Hartree–Fock limit.
3. T.E.H. Walker and W.G. Richards, *Symp. Faraday Soc.* **2**, 64 (1968). Calculation of spin–orbit coupling constant.
4. P.E. Cade, R.F.W. Bader, W.H. Henneker, and J. Keaveny, *J. chem. Phys.* **50**, 5313 (1969). Charge distribution.

SiH^+

State	Internuclear distance	Basis set	Energy	Wave function	Reference
$X\,^1\Sigma^+$	2·874	Ext. STO	−289·1656		1

REFERENCE

1. P.E. Cade and W.M. Huo, *J. chem. Phys.* **47**, 649 (1967). Near Hartree–Fock limit.

SiH^-

State	Internuclear distance	Basis set	Energy	Wave function	Reference
$X\,^3\Sigma^-$	2·861	Ext. STO	−289·4598	✓	1

Property	*Reference*
Orbital energies	1

REFERENCE

1. P.E. Cade, *Proc. phys. Soc.* **96**, 842 (1967).

SiO

State	Internuclear distance	Basis set	Energy	Wave function	Reference
$^1\Sigma^+$	2·75	Ext. STO	–363·8523	✓	1

Properties	*Reference*
Potential curve	1
Orbital energies	1

REFERENCE

1. A.D. McLean and M. Yoshimine, Supp. to *I.B.M. Jl Res. Dev.* (1967). Near Hartree–Fock limit.

SrO

State	Internuclear distance	Basis set	Energy	Wave function	Reference
$^1\Sigma^+$	3·6283	Ext. STO	–3206·2311		1
	3·6283	Ext. STO	–3206·2311	✓	2
	3·6283	Ext. STO	–3206·2311	✓	3

Properties	*Reference*
Orbital energies	2
Potential curve	2
Spectroscopic constants	3
Dipole moment	1, 3
Quadrupole coupling	3
Magnetic properties	3

REFERENCES

1. M. Yoshimine and A.D. McLean, *Int. J. Quantum Chem.* **1S**, 313 (1967).

2. A.D. McLean and M. Yoshimine, Supp. to *I.B.M. Jl Res. Dev.* (1967).

3. M. Yoshimine, *J. phys. Soc. Japan* **25**, 1100 (1968).
This wave function is near the Hartree–Fock limit.

TiN

State	Internuclear distance	Basis set	Energy	Wave function	Reference
$X\,^2\Sigma^+$	2·84	Small STO			1

Properties	*Reference*
Orbital energies	1
Excited states	1
Spectroscopic constants	1
Dipole moment	1

REFERENCE

1. K.D. Carlson, C.R. Claydon, and C.M. Moser, *J. chem. Phys.* **46**, 4693 (1967).
Discussion of excited states.

TiO

State	Internuclear distance	Basis set	Energy	Wave function	Reference
$X\,^3\Delta$	3·0618	Min. STO	−921·5418	✓	2
	2·91	Ext. STO	−922·4933	✓	3
$^1\Sigma^+$	3·0618	Min. STO	−921·4959	✓	2
	2·91	Ext. STO	−922·4373	✓	3

Properties	*References*
Orbital energies	2, 3
Potential curve	2, 3
Spectroscopic constants	2, 3
Excited states	1, 2, 3
Dipole moment	2
Charge density	4

REFERENCES

1. K.D. Carlson and C.M. Moser, *J. phys. Chem., Ithaca,* **67**, 2644 (1963).
Min. STO used for low-lying states.

2. K.D. Carlson and R.K. Nesbet, *J. chem. Phys.* **41**, 1051 (1964).
Discussion of ground state.

3. K.D. Carlson and C.M. Moser, *J. chem. Phys.* **46**, 35 (1967). Discussion of ground state.

4. I. Cohen and K.D. Carlson, *J. phys. Chem., Ithaca*, **73**, 1356 (1969). Charge densities.

VO

State	Internuclear distance	Basis set	Energy	Wave function	Reference
$X\,^4\Sigma^-$	2·91	Ext. STO	−1015·8917	✓	1
$^2\Delta$	2·91	Ext. STO	−1015·7627		1

Properties	*References*
Spectroscopic constants	1
Orbital energies	1
Excited states	1
Dipole moment	1

REFERENCE

1. K.D. Carlson and C.M. Moser, *J. chem. Phys.* **44**, 3259 (1966). Prediction of ground state.

H_3

State	Geometry	Basis set	Energy	Wave function	Reference
	R(H–H)				
$^2\Sigma_u^+$	1·73	Ext. GTO	−1·5930		5
	1·7	Floating 1s GTO	−1·5940		14
	1·8	One centre STO + CI	−1·6121		8
	1·7	One centre STO	−1·63074		11
	1·771	STO + CI	−1·63792		17
		STO + CI	−1·6473	✓	19
	1·8	Pseudo-natural orbitals	−1·6493	✓	6, 15
	1·8	STO + CI	−1·6521		13
	1·75	Soln. of Schrödinger equation.	−1·6551		9, 10

Properties	*References*
Potential surface	5, 6, 7, 9, 10, 12, 13, 14, 15 17, 19

REFERENCES

1. J.O. Hirschfelder, H. Eyring, and N. Rosen, *J. chem. Phys.* **4**, 121 (1936).
   MO study. 1s orbitals with effective charge.
2. J.M. Walsh and F.A. Matsen, *J. chem. Phys.* **19**, 526 (1951).
   Binding energy by M.O. calc.
3. G.E. Kimball and J.G. Trulis, *J. chem. Phys.* **28**, 493 (1958).
   Molecular orbital calc.
4. H. Aroeste and W.J. Jameson, *J. chem. Phys.* **30**, 372 (1959).
   Short-range interactions of $H + H_2$.
5. M. Krauss, *J. Res. natn. Bur. Stand.* **68A**, 635 (1964).
6. G. Edmiston and M. Krauss, *J. chem. Phys.* **42**, 1119 (1965).
7. H. Conroy and B.L. Bruner, *J. chem. Phys.* **42**, 4047 (1965).
8. J.P. Considine and E.F. Hayes, *J. chem. Phys.* **46**, 1119 (1967).
   Includes 52-term CI.
9. H. Conroy, *J. chem. Phys.* **47**, 912 (1967).
10. H. Conroy and B.L. Bruner, *J. chem. Phys.* **47**, 921 (1967).
11. E.F. Hayes and R.G. Parr, *J. chem. Phys.* **47**, 3961 (1967).
12. H.H. Michels and F.E. Harris, *J. chem. Phys.* **48**, 2371 (1968).
    Long-range interactions.
13. I. Shavitt, R.M. Stevens, F.L. Minn, and M. Karplus, *J. chem. Phys.* **48**, 2700 (1968).
14. H.E. Schwartz and L.J. Schaad, *J. chem. Phys.* **48**, 4709 (1968).
    Includes work on H_4^+ and H_4.
15. C. Edmiston and M. Krauss, *J. chem. Phys.* **49**, 192 (1968).
    Pseudo-natural orbitals.
16. R.N. Porter, R.M. Stevens, and M. Karplus, *J. chem. Phys.* **49**, 5163 (1968).
    Discussion of Jahn–Teller distortion.
17. A. Riera and J.W. Linnett, *Theor. Chim. Acta* **15**, 181 (1969).
18. J.W. Linnett and A. Riera, *Theor. Chim. Acta* **15**, 196 (1969).
    Analysis of wave function of ref. 17
19. E. Giacometti, G.F. Majorino, E. Rusconi, and M. Simonetta, *Int. J. Quantum Chem.* **3**, 45 (1969).

H_3^+

State	Geometry	Basis set	Energy	Wave function	Reference
	R(H–H)				
$^1\Sigma_g^+$	1·8	One centre STO	−1·24657		15
	1·55	Elliptical	−1·2460	✓	10
		GTO. 2 electron	−1·2729		17
$^1A_1$	1·68	Elliptical	−1·27264	✓	10
	1·65	GTO	−1·2772		21
	1·6229	Single centre STO	1·2863		13
	1·68	Ext. GTO	−1·2984		11
	1·6405	Floating 1s GTO	−1·29993		16
	1·6504	+ CI	−1·33764		16
	1·6575	STO + CI	−1·30432		6
	1·66	V.B. GTO	−1·3185		12
	1·6575	STO + CI	−1·3326	✓	8
		GTO. 2 electron	−1·3359		17
	1·68	2 electron	−1·357		7, 9
$^1A_1$, $^1B_1$, and $^1B_2$		Soln. of molecular Schrödinger equation			23

Properties	*References*
Potential surface	6, 7, 8, 9
Excited states	8, 23
Spectroscopic constants	8, 12, 22
Thermodynamic properties	16
Proton affinities	21

REFERENCES

1. H.W. Massey, *Proc. Camb. phil. Soc.* **27**, 451 (1931).
Valence bond.

2. C.A. Coulson, *Proc. Camb. phil. Soc.* **31**, 244 (1935).
Molecular orbital calculation.

3. J.O. Hirschfelder, H. Eyring, and N. Rosen, *J. chem. Phys.* **4**, 130 (1936).
M.O. calculation.

4. R.G. Pearson, *J. chem. Phys.* **16**, 502 (1948).
Molecular orbital calculation.

5. R.S. Barber, J.C. Giddings, and H. Eyring, *J. chem. Phys.* **23**, 344 (1955).
Molecular orbital calculation on linear molecule.

6. R.E. Christoffersen, S. Hagstrum, and F. Prosser, *J. chem. Phys.* **40**, 236 (1964).
Valence bond.

7. H. Conroy, *J. chem. Phys.* **40**, 603 (1964).
Soln. of two-electron Schrödinger equation.

8. R.E. Christoffersen, *J. chem. Phys.* **41**, 960 (1964).

9. H. Conroy, *J. chem. Phys.* **41**, 1341 (1964).
Minimizes energy in two-electron problem.

10. J.R. Hoyland, *J. chem. Phys.* **41**, 1370 (1964).
Two-centre wave function.

11. W.A. Lester and M. Krauss, *J. chem. Phys.* **44**, 207 (1966).
Correlated wave function.

12. A.G. Pearson, R.D. Poshusta, and J.C. Browne, *J. chem. Phys.* **44**, 1815 (1966).
Valence bond.

13. B.D. Joshi, *J. chem. Phys.* **44**, 3627 (1966).
Single centre.

14. H. Conroy and B.L. Bruner, *Theoretical Chem. Preprint No. 22 Mellon Inst.* (1960).
Minimizes energy variance.

15. J. Considine and E.F. Hayes, *J. chem. Phys.* **46**, 1119 (1967).
Single centre + CI. 21-term CI function.

16. M.E. Schwartz and L.J. Schaad, *J. chem. Phys.* **47**, 5325 (1967).

17. W. Kutzelnigg, R. Ahlrichs, I. Labib–Ishander, and W.A. Bingel, *Chem. Phys. Lett.* **1**, 447 (1967).
Discussion of correlation energy.

18. A.A. Wu and F.O. Ellison, *J. chem. Phys.* **48**, 1491 (1968).
Diatomics in molecules.

19. R.E. Christoffersen and H. Shull, *J. chem. Phys.* **48**, 1790 (1968).
Nature of bonding.

20. A.A. Wu and F.O. Ellison, *J. chem. Phys.* **48**, 5032 (1968).
Diatomics in molecules.

21. A.C. Hopkinson, N.K. Holbrook, K. Yates, and I.G. Csizmadia, *J. chem. Phys.* **49**, 3596 (1968).
Various basis sets used.

22. R.D. Poshusta, J.A. Haugen, and D.F. Zetik, *J. chem. Phys.* **51**, 3343 (1969).

23. H. Conroy, *J. chem. Phys.* **51**, 3979 (1969).
Soln. of molecular Schrödinger equation also includes H_3^{++}.

24. H. Preuss and R. Janoschek, *J. molec. Struct.* **3**, 423 (1969).
Reaction surface for $H_3^+ + He = HeH^+ + H_2$.

H_3^-

State	Geometry	Basis set	Energy	Wave function	Reference
	R(H–H)				
$^1\Sigma_g^+$	2·2098	STO + CI	−1·581076		2
	2·02	STO + CI	−1·66742	✓	3

REFERENCES

1. R.S. Barker, H. Eyring, D.A. Baker, and C.J. Thorne, *J. chem. Phys.* **23**, 1381 (1955). Comparison of methods.

2. A. Macias, *J. chem. Phys.* **48**, 3464 (1968).

3. A. Macias, *J. chem. Phys.* **49**, 2198 (1968).

HeH_2

Property	*Referencees*
Potential curve	1, 2

REFERENCES

1. C.S. Roberts, *Phys. Rev.* **131**, 203 (1963).
V.B. calculation with min. STO on $^1A_1$. Energy at large values of R(He–H).

2. M. Krauss and F.H. Mies, *J. chem. Phys.* **42**, 2703 (1965).
Ext. GTO and STO basis for $^1A_1$ state. Approaches Hartree–Fock limit.

HeH_2^+

Properties	*Reference*
Spectroscopic constants	1
Binding energy	1

REFERENCE

1. R.D. Poshusta, J.A. Haugen, and D.F. Zetik, *J. chem. Phys.* **51**, 3343 (1969).
V.B. calculation with GTO on ground state. R(H–He) = 2·58, R(H–He) = 1·46.

NeH_2

State	Geometry	Basis set	Energy	Wave function	Reference
	R(Ne–H)				
$^1\Sigma_g^+$		GTO	repulsive		1
$^1\Sigma_u^+$	3·0	GTO	−129·0155		1
$^3\Sigma_g^+$	3·5	GTO	−128·7902		1
$^3\Sigma_u^-$		GTO	repulsive		1
$^3\Sigma_g^-$	3·0	GTO	−127·5984		1

Property	*Reference*
Potential curves	1

REFERENCE

1. R. Kapal, *J. chem. Phys.* **46**, 2317 (1967).
Valence bond and Gaussian lobe calculations.

LiH_2

Property	*Reference*
Energy surface	1

REFERENCE

1. M. Krauss, *J. Res. natn. Bur. Stand.* **72A**, 553 (1968).
Energy surface for Li/H_2^+ reaction.

LiH_2^+

Properties	*Reference*
Spectroscopic constants	1
Binding energy	1

1. R.D. Poshusta, J.A. Haugen, and D.F. Zetik, *J. chem. Phys.* **51**, 3343 (1969).
V.B. with GTO basis for $^1A_1$ state.

LiH_2^-

State	Geometry	Basis set	Energy	Wave function	Reference
	R(Li–H)				
$^1\Sigma_g$	3·514	FSGO	–7·026		2
	3·50	contr. GTO			1
	3·49	GTO	–8·531		3

Properties	*References*
Hypersurface	1, 3
Dissociation energy	1

REFERENCES

1. H. Preuss and G. Diercksen, *Int. J. Quantum Chem.* **1**, 631 (1967).
2. A.A. Frost, *J. phys. Chem., Ithaca,* **72**, 1289 (1968).
Floating spherical gaussians
3. H. Preuss and R. Janoschek, *J. molec. Struct.* **3**, 423 (1969).

BeH_2

State	Geometry	Basis set	Energy	Wave function	Reference
	R(Be–H)				
$^1\Sigma_g^+$	2·54	GTO lobe	–15·7123		1
	2·5	STO + CI	–15·7202		4

Property	*References*
Potential curve	1, 4

REFERENCES

1. S.D. Peyerimhoff, R.J. Buenker, and L.C. Allen, *J. chem. Phys.* **45**, 734 (1966).
2. M.C. Goldberg and J.R. Riter, *J. phys. Chem., Ithaca,* **71**, 3111 (1967).
Predicts Hartree–Fock limit.
3. A.A. Frost, B.H. Prentice, and R.A. Rouse, *J. Am. chem. Soc.* **89**, 3064 (1967).
Floating spherical gaussian calculation.

4. F. E. Harris and H.H. Michels, *Int. J. Quantum Chem.* **1S**, 329 (1967). Valence bond calculation.

5. R. Ahlrichs and W. Kutzelnigg, *Theor. Chim. Acta,* **10**, 377 (1968). Natural orbitals from gaussian basis. Estimate of correlation energy.

6. A.A. Wu and F.O. Ellison, *J. chem. Phys.* **48**, 727 (1968). Diatomics in molecules calculation.

BeH_2^+

REFERENCE

1. R.D. Poshusta, J.A. Haugen, and D.F. Zetik, *J. chem. Phys.* **51**, 3343 (1969).

BH_2

State	Geometry	Basis set	Energy	Wave function	Reference
	R(B–H)				
$^1\Sigma_g^+$	2·54	Gaussian lobe	– 25·71227	√	1

Properties	*Reference*
Orbital energy and change with angle	1

REFERENCE

1. S.D. Peyerimhoff, R.J. Buenker, and L.C. Allen, *J. chem. Phys.* **45**, 734 (1966).

BH_2^+

State	Geometry	Basis set	Energy	Wave function	Reference
	R(B–H)				
$^1\Sigma_g^+$	2·292	Floating GTO	–21·563		3
	2·35	Gaussian lobe	–25·4112		2
	2·45	Gaussian lobe	–25·4363	√	1

Properties	*References*
Orbital energy and change with angle	1, 2

REFERENCES

1. S.D. Peyerimhoff, R.J. Buenker, and L.C. Allen, *J. chem. Phys.* **45**, 734 (1966).

2. S.D. Peyerimhoff, R.J. Buenker, and J.L. Whitten, *J. chem. Phys.* **46**, 1707 (1967).

3. A.A. Frost, *J. phys. Chem., Ithaca*, **72**, 1289 (1968). Floating spherical gaussian.

BH_2^-

State	Geometry R(B–H)	∠	Basis set	Energy	Wave function	Reference
$^1A_1$	2·45	105°	Gaussian lobe	–25·6511		2
	2·35	105°	Gaussian lobe	–25·68233	√	1

Property	*References*
Orbital energy	1, 2

REFERENCES

1. S.D. Peyerimhoff, R.J. Buenker, and L.C. Allen, *J. chem. Phys.* **45**, 734 (1966).

2. S.D. Peyerimhoff, R.J. Buenker, and J.L. Whitten, *J. chem. Phys.* **46**, 1707 (1967).

CH_2

State	Geometry R(C–H)	∠	Basis set	Energy	Wave function	Reference
$^3\Sigma_g^-$	1·95		GTO	–38·8936	√	2
$^3B_1$	2·00	140°	V.B. (GTO)	–38·9151	√	3
	2·11	129°	Min. STO + CI	–38·904		1
$^1A_1$	2·1	105°	GTO	–38·8026	√	2
	2·00	100°	V.B. (GTO)	–38·8643	√	3
	2·21	90°	Min. STO + CI	–38·865		1

State	Geometry		Basis set	Energy	Wave function	Reference
	R(C–H)	∠				
$^1B_1$	2·11	132°	Min. STO + CI	–38·808		1
	2·00	140°	V.B. (GTO)	–38·8322	✓	3

Properties	*References*
Spectroscopic constants	1, 3
Potential curve [fn(HĈH)]	1, 3
Orbital energies	2
Magnetic properties	3
Dipole moment	3
Quadrupole moment	3
Term values	3

REFERENCES

1. J.M. Foster and S.F. Boys, *Rev. mod. Phys.* **32**, 305 (1960).
2. M. Krauss, *J. Res. natn. Bur. Stand.* **68A**, 635 (1964).
3. J.F. Harrison and L.C. Allen, *J. Am. chem. Soc.* **91**, 807 (1969).

NH_2

State	Geometry	Basis set	Energy	Wave function	Reference
$^2B_1$	R = 1·9082 ∠ = 120°	one centre STO	–55·3561	✓	1
		GTO	–55·5249		2

REFERENCES

1. A.L.H. Chung, *J. chem. Phys.* **46**, 3144 (1966).
CI calculations also performed.
2. T.A. Claxton, D. McWilliams, and N.A. Smith, *Chem. Phys. Lett.* **4**, 505 (1969).
Unrestricted Hartree–Fock.
3. W. Mayer, *J. chem. Phys.* **51**, 5149 (1969).

NH_2^+

State	Geometry	Basis set	Energy	Wave function	Reference
$^1A_1$	R(N–H) 2·05 ∠ = 120°	GTO lobe	–55·0853	√	1, 2

Properties	*References*
Potential curve	1, 2
Orbital energies	1, 2

REFERENCES

1. S.D. Peyerimhoff, R.J. Buenker, and L.C. Allen, *J. chem. Phys.* **45**, 734 (1966).
Discusses Walsh diagram.

2. S.D. Peyerimhoff, R.J. Buenker, and J.L. Whitten, *J. chem. Phys.* **46**, 1707 (1967).

NH_2^-

State	Geometry R(N–H)	Geometry ∠	Basis set	Energy	Wave function	Reference
$^1A_1$	1·964	86·4°	FSGO	–46·791		2
	1·958	107°	970	–55·5009		1

Property	*Reference*
Proton affinity	1

REFERENCES

1. A.C. Hopkinson, N.K. Holbrook, K. Yates, and I.G. Csizmadia, *J. chem. Phys.* **49**, 3596 (1968).
Various basis sets used.

2. A.A. Frost, *J. phys. Chem., Ithaca*, **72**, 1289 (1968).
Floating spherical gaussian orbitals.

H_2O

State	Geometry R(O–H)	∠	Basis set	Energy	Wave function	Reference
$X\,^1A_1$	1·663		FSGO	−64·290		21, 27
	1·77		One centre	−74·11		4
	1·81	105°	One centre	−75·00		6
	1·81	104·4°	GTO	−75·002		33
	1·89	110°	Min. STO	−75·6812		35
	1·8103	100°	Min. STO	−75·6923		29
			Min. STO	−75·7201	✓	34
			Min. STO	−75·761		8
	1·813		GTO	−75·767		36
	1·902	98°	Min. STO	−75·7684		18
	1·8104	105°	V.B. + CI	−75·7763		13
	1·80	105°	GTO	−75·8453	✓	11
	1·8103	120°	STO	−75·8467	✓	2
	1·8236	104·5°	STO	−75·893		19
	1·81	105°	GTO	−75·9117		30
	1·814	107°	One centre	−75·9224		9, 10
	1·81	105°	G. lobe	−75·9734		32
	1·811	104·25°	STO	−76·0047	✓	23
			GTO	−76·0127		28
	1·809	104·5°	GTO	−76·0179		25
	1·8	105°	GTO	−76·034		15
	1·8	105°	GTO	−76·0419		31
	1·8	105°	Ext. GTO	−76·0421		12
	1·8	105°	Ext. GTO	−76·0596	✓	22
$^3B_1$	1·81	105°	GTO	−75·6435		30
	1·80	105°	GTO	−75·8153		31
$A\,^1B_1$	1·81	105°	GTO	−75·6214		30
	1·80	105°	GTO	−75·8029		31
$^3A_1$	1·81	105°	GTO	−75·5813		30
$^1B_1$	1·81	105°	GTO	−75·5454		30
$^{1,3}A''$			GTO	repulsive		31
$^1\Sigma_g^+$	3·06		FSGO	−64·203		21

Properties	*References*
Dipole moment	2, 8, 9, 10 12, 19, 22, 23, 24 29, 31, 32, 33, 34, 35, 36

Properties	*References*
Orbital energies	2, 4, 12, 22, 23, 30
X-ray form factor	3, 5
Charge density	5, 18, 34
Magnetic properties	7, 22, 23, 24
Localized orbitals	8, 20, 32
Polarizabilities	16, 26
Quadrupole coupling	17, 22, 24
Force constants	18, 24, 35
Proton affinity	25
Excited states	28, 30, 31
Atomic population	29
Oscillator strengths	31
Ionization potential	35, 36

REFERENCES

1. A.S. Coolidge, *Phys. rev.* **42**, 189 (1932).
Min. STO for V.B. calculation.

2. F.O. Ellison and H. Shull, *J. chem. Phys.* **23**, 2348 (1955).

3. K.E. Banyard and N.H. March, *Acta cystallogr.* **9**, 385 (1956).
X-ray form factor calculated.

4. K. Funabaski and J.L. Magee, *J. chem. Phys.* **26**, 407 (1957).

5. K.E. Banyard and N.H. March, *J. chem. Phys.* **26**, 1416 (1957).

6. K.E. Banyard and N.H. March, *J. chem. Phys.* **27**, 977 (1957).
One-centre calculations.

7. K.E. Banyard, *J. chem. Phys.* **33**, 832 (1960).

8. R. McWeeny and K.A. Ohno, *Proc. R. Soc.* **A255**, 367 (1960).
Discusses bond approximation.

9. R. Moccia, *J. chem. Phys.* **37**, 910 (1962).

10. R. Moccia, *J. chem. Phys.* **40**, 2186 (1964).
One centre ext. STO.

11. M. Krauss, *J. Res. natn. Bur. Stand.* **68A**, 635 (1964).

12. J.W. Moskowitz and M.C. Harrison, *J. chem. Phys.* **43**, 3550 (1965).
Ext. GTO. Walsh diagrams discussed.

13. B. Hockel, D. Hamel, and K. Ruchelhausen, *Z. Naturf.* **20a**, 26 (1965).
Valence bond with CI.

14. L.C. Snyder, *J. chem. Phys.* **46**, 3602 (1967).
Estimate of Hartree–Fock limit.

15. C.D. Ritchie and H.F. King, *J. chem. Phys.* **47**, 564 (1967).
Estimate of Hartree–Fock limit.

16. G.P. Arrighini, M. Maestro, and R. Moccia, *Chem. Phys. Lett.* **1**, 242 (1967).
Electric polarizability.

17. P. Pykko, *Proc. phys. Soc.* **92**, 841 (1967).
Deutron quadrupole coupling constant.

18. D. Hamel, *Z. Naturf.* **22a**, 176 (1967).

19. D. Hager, E. Hess, and L. Zühicke, *Z. Naturf.* **22a**, 1282 (1967).
CI also carried out.

20. J.G. Stamper and N. Trinajstic, *J. chem. Soc.* 782 (1967).
Localized orbitals.

21. A.A. Frost, B.H. Prentice, and R.A. Rouse, *J. Am. chem. Soc.* **89**, 3064 (1967).
Floating spherical gaussians.

22. D. Neumann and J.W. Moskowitz, *J. chem. Phys.* **49**, 2056 (1968).
Near Hartree–Fock limit.

23. S. Aung, R.M. Pitzer, and S.I. Chan. *J. chem. Phys.* **49**, 2071 (1968).
Different basis sets compared.

24. C.W. Kern and R.L. Matcha, *J. chem. Phys.* **49**, 2081 (1968).
One-electron properties corrected for nuclear motion.

25. A.C. Hopkinson, N.K. Holbrook, K. Yates, and I.G. Csizmadia, *J. chem. Phys.* **49**, 3596 (1968).
Proton affinity calculated.

26. G.P. Arrighini, M. Maestro, and R. Moccia, *Symp. Faraday Soc.* **2**, 48 (1968).
Hyperpolarizability.

27. A.A. Frost, *J. phys. Chem., Ithaca,* **72**, 1289 (1968).
Floating spherical gaussians.

28. W.J. Hunt and W.A. Goddard, *Chem. Phys. Lett.* **3**, 414 (1969).
Excited states by virtual orbitals.

29. M. Klessinger, *Chem. Phys. Lett.* **4**, 144 (1969).
Self consistent group calculation.

30. J.A. Horsley and W.H. Fink, *J. chem. Phys.* **50**, 750 (1969).
Uses excited states to discuss photodissociation of H_2O.

31. K.J. Miller, S.R. Miekzarek, and M. Krauss, *J. chem. Phys.* **51**, 26 (1969).
Dissociation discussed.

32. J.D. Petke and J.L. Whitten, *J. chem. Phys.* **57**, 3166 (1969). Orbital hybridization discussed.

33. I.H. Hillier and V.R. Saunders, *Chem. Phys. Lett.* **4**, 163 (1969).

34. E. Switkes, R.M. Stevens, and W.N. Lipscomb, *J. chem. Phys.* **51**, 5229 (1969).

35. J. Andriessen, *Chem. Phys. Lett.* **3**, 257 (1969).

36. H. Preuss and R. Janoschek, *J. molec. Struct.* **3**, 423 (1969).

H_2O^+

State	Geometry R(O–H)	∠	Basic set	Energy	Wave function	Reference
$^2B_1$	1.9	115°	GTO	–75.4465		1
	1.8	105°	GTO	–75.62799		2
$^2A_1$	1.8	105°	GTO	–75.5352		2
$^2B_2$	1.8	105°	GTO	–75.3830		2

REFERENCES

1. M. Krauss, *J. Res. natn. Bur. Stand.* **68A**, 635 (1964).

2. K.J. Miller, S.R. Miekzarek, and M. Krauss, *J. chem. Phys.* **51**, 26 (1969).

H_2S

State	Geometry R(S–H)	∠	Basis set	Energy	Wave function	Reference
$^1A_1$	2.5228	92.25°	Ext. GTO	–381.03894		5
	2.523	92.5°	Ext. GTO	–381.0391		6
	2.510	92.2°	Ext. GTO	–394.516		8
	2.523	92.25°	Ext. GTO	–396.9005		9
	2.509	89°	One centre	–397.5888		1
	2.509	89°	One centre	–397.5891		2
	2.509	92.2°	Min. STO	–397.8415	✓	7

Properties	*References*
Dipole moment	1, 2, 5, 8
Deutron quadrupole coupling constant	4

Properties	*References*
Orbital energies	5, 7, 9
Charge density	5, 9
Proton affinity	6
Ionization potential	8

REFERENCES

1. R. Moccia, *J. chem. Phys.* **37**, 910 (1962).
2. R. Moccia, *J. chem. Phys.* **40**, 2186 (1964).
3. D.B. Cook and P. Palmieri, *Chem. Phys. Lett.* **3**, 219 (1969). Mixed STO and GTO calculation.
4. P. Pyykko, *Proc. phys. Soc.* **92**, 841 (1967).
5. A. Rauk and I.G. Csizmadia, *Can. J. Chem.* **46**, 1205 (1968).
6. A.C. Hopkinson, N.K. Holbrook, K. Yates and I.G. Csizmadia, *J. chem. Phys.* **49**, 3596 (1968).
7. F.P. Boer and W.N. Lipscomb, *J. chem. Phys.* **50**, 989 (1969).
8. I.H. Hillier and V.R. Saunders, *Chem. Phys. Lett.* **4**, 163 (1969).
9. M.E. Schwartz, *J. chem. Phys.* **51**, 182 (1969).

H_2S^+

State	Geometry	Basis set	Energy	Wave function	Reference
$^2B_1$		GTO	–398.2335		

Properties	*References*
Hyperfine constants	1

REFERENCE

1. T.A. Claxton, D. McWilliams, and N.A. Smith, *Chem. Phys. Lett.* **4**, 505 (1969). Unrestricted Hartree-Fock.

H_2S^-

State	Geometry	Basis set	Energy	Wave function	Reference
$^2A_1$		GTO	–398.3233		1

Properties	*References*
Hyperfine constants	1

REFERENCE

1. T.A. Claxton, D. McWilliams, and N.A. Smith, *Chem. Phys. Lett.* **4**, 505 (1969).
Unrestricted Hartree–Fock.

H_2F^+

State	Geometry	Basis set	Energy	Wave function	Reference
	R(F–H) ∠				
$^1A_1$	1.733 105°	GTO	–100.2082		1

Property	*Reference*
Proton affinity	1

REFERENCES

1. A.C. Hopkinson, N.K. Holbrook, K. Yates and I.G. Csizmadia, *J. chem. Phys.* **49**, 3596 (1968).
Various basis sets used.
2. A.A. Frost, *J. phys. Chem., Ithaca,* **72**, 1289 (1968).
Floating spherical gaussian calculation.

H_2F^-

State	Geometry	Basis set	Energy	Wave function	Reference
$^1\Sigma^+$	R(F–H) 3.60 R(H–H) 1.45	Ext. GTO	–100.586		1

Property	*Reference*
Potential curve	1

REFERENCE

1. C.D. Ritchie and H.F. King, *J. Am. chem. Soc.* **88**, 1069 (1966).

He_2H^+

State	Geometry	Basis set	Energy	Wave function	Reference
$^1\Sigma$	R(He–H) 1.775	GTO	–5.7930		2, 3

Properties	*References*
Spectroscopic constants	1, 2
Binding energy	1

REFERENCES

1. R.D. Poshusta, J.A. Haugen, and D.F. Zetik, *J. chem. Phys.* **51**, 3343 (1969).
   V.B. calculation with GTO. R(He–H) = 1.70.
2. R. Swanstrom, R. Janoschek and H. Preuss, *Int. J. Quantum Chem.* **3**, 115 (1969).
3. H. Preuss and R. Janoschek, *J. molec. Struct.* **3**, 423 (1969).
   Includes excited states.

HeLiH

State	Geometry	Basis set	Energy	Wave function	Reference
$^1\Sigma^+$	R(Li–H) 3.015 R(Li–He) 4.00	Ext. GTO	–10.81876		1

Properties	*Reference*
Dipole moment	1
Orbital energies	1
Potential curve	1
Charge density	1
Atomic population	1

REFERENCE

1. J.J. Kaufman and L.M. Sachs, *J. chem. Phys.* **51**, 2992 (1969).
   Complete potential curves calculated.

$HeLiH^+$

State	Geometry	Basis set	Energy	Wave function	Reference
$^2\Sigma^+$	R(Li–He) 3.06 R(Li–H) 3.85	VB/GTO			1

Properties	*Reference*
Spectroscopic constants	1
Binding energy	1

REFERENCE

1. R.D. Poshusta, J.A. Haugen, and D.F. Zelik, *J. chem. Phys.* **51**, 3343 (1969).
Valence bond calculation.

Li_2H^+

State	Geometry	Basis set	Energy	Wave function	Reference
$^1\Sigma_g^+$	R(Li–H) 3.1	Contr. GTO	–15.289		1

Properties	*Reference*
Potential curve	1
Ionization potential	1

REFERENCE

1. G. Diercksen and H. Preuss, *Int. J. Quantum Chem.* **1**, 637 (1967).

Li_2H^-

State	Geometry	Basis set	Energy	Wave function	Reference
$^1\Sigma^+$	R(Li–H) 3.41	Contr. GTO	–15.3874		1, 2

Property	*Reference*
Spectroscopic constants	1

REFERENCES

1. H. Preuss and G. Diercksen, *Int. J. Quantum Chem.* **1**, 641 (1967).
2. H. Preuss and R. Janoschek, *J. molec. Struct.* **3**, 423 (1969).

Be_2H^+

State	Geometry	Basis set	Energy	Wave function	Reference
	R(Be–H)				
$^1\Sigma_g^+$	3.039	GTO	–29.383		1
$^1\Sigma_u^-$		GTO			1
$^1\Pi_u$		GTO			1

Properties	*Reference*
Vibration frequencies	1
Ionization potential	1

REFERENCE

1. H. Preuss and R. Janoschek, *J. molec. Struct.* **3**, 423 (1969).

Be_2H^-

REFERENCE

1. H. Preuss and R. Janoschek, *J. molec. Struct.* **3**, 423 (1969). Unstable molecule.

LiOH

State	Geometry	Basis set	Energy	Wave function	Reference
$^1\Sigma$	R(Li–O) 3.0236				
	R(O–H) 1.823				
	linear	Ext. STO	–82.9092		1
	$\angle = 140°$		–82.9018		1

Properties	*Reference*
Charge density	1
Orbital energy	1
Potential curves	1

REFERENCE

1. R.J. Buenker and S.D. Peyerimhoff, *J. chem. Phys.* **45**, 3682 (1966). Discussion of Walsh's rules.

HCN

State	Geometry		Basis set	Energy	Wave function	References
	R(C–N)	R(C–H)				
$^1\Sigma^+$	2.187	2.0	Min. STO	–92.5471		11
	2.187	2.00	Min. STO	–92.5474	√	2
	2.187	2.006	Min. STO	–92.5903	√	5
	2.187	2.0	STO	–92.6577		15
	2.19	2.01	GTO	–92.6624		12
	2.187	2.00	GTO	–92.7140		8
	2.1775	2.0	Gaussian lobe	–92.8287	√	7
	2.1791	2.0143	Ext. STO	–92.9089		10
	2.179	2.014	Ext. STO	–92.9147	√	9

Properties	*References*
Atomic population	1, 5, 11, 15
Charge density	3, 5, 11
Quadrupole coupling	4, 7, 11, 13
Localized orbitals	6
Potential curve	7, 8, 9
Dipole moment	7, 10, 14, 15
Orbital energies	8, 9, 10

REFERENCES

1. E. Clementi and H. Clementi, *J. chem. Phys.* **36**, 2824 (1962). Atomic population.

2. A.D. McLean, *J. chem. Phys.* **37**, 627 (1962). Min. STO.

3. L. Burnelle, *Theor. Chim. Acta* **2**, 177 (1964). Charge density.

4. C.W. Kern and M. Karplus, *J. chem. Phys.* **42**, 1062 (1965). Quadrupole coupling constant.

5. W.E. Palke and W.N. Lipscomb, *J. Am. chem. Soc.* **88**, 2384 (1966). Population analysis.

6. J.G. Stamper and N. Trinajstic, *J. chem. Soc.* **A782** (1967).
Localized orbitals.

7. D.C. Pan and L.C. Allen, *J. chem. Phys.* **46**, 1797 (1967).
Discussion of Walsh's diagrams.

8. J.B. Moffat and R.J. Collens, *Can. J. Chem.* **45**, 655 (1967).

9. A.D. McLean and M. Yoshimine, Supp. to *I.B.M. J. Res. Dev.* (1967).
Near Hartree–Fock limit.

10. M. Yoshimine and A.D. McLean, *Int. J. Quantum Chem.* **1S**, 313 (1967).
Extended STO.

11. R. Bonaccorsi, C. Petrongolo, E. Scrocco, and J. Tomasi, *J. chem. Phys.* **48**, 1500 (1967).

12. A.C. Hopkinson, N.K. Holbrook, K. Yates, and I.G. Csizmadia, *J. chem. Phys.* **49**, 3596 (1968).

13. R. Bonaccorsi, E. Scrocco, and J. Tomasi, *J. chem. Phys.* **50**, 2940 (1969).
Electric field gradient.

14. J.B. Moffat, *Chem. Communs.* 789 (1966).
Gaussian basis, $E = -92.257$.

15. E. Switkes, R.M. Stevens, and W.W. Lipscomb, *J. chem. Phys.* **51**, 5229 (1969).

HCO

Property	*Reference*
Hyperfine constants	1

REFERENCE

1. A. Hinchcliffe and D.B. Cook, *Chem. Phys. Lett.* **1**, 217 (1967).
Uses wave function based on one for formaldehyde.

HCO^+

State	Geometry	Basis set	Energy	Wave function	Reference
$^1\Sigma^+$	R(C–O) 2.129 R(C–H) 2.076	GTO	−112.6593		1

REFERENCE

1. A.C. Hopkinson, N.K. Holbrook, K. Yates, and I.G. Csizmadia, *J. chem. Phys.* **49**, 3596 (1958).
Proton affinity of CO calculated.

HNO

State	Geometry	Basis set	Energy	Wave function	Reference
$X\,^1A_1$	R(N–H) 1.962 R(N–O) 2.496 $\angle = 110°$	GTO	–129.623		1

Property	*Reference*
Potential curve	1

REFERENCE

1. A.W. Salotto and L. Burnelle, *Chem. Phys. Lett.* **3**, 80 (1969).
Projected UHF wave function.

FOH

State	Geometry	Basis set	Energy	Wave function	Reference
$^1A'$	R(F–H) 2.065 R(O–H) 1.833 $\angle = 100°$	STO	–174.7196		1

Properties	*Reference*
Charge density	1
Potential curves	1
Orbital energies	1

REFERENCE

1. R.J. Buenker and S.D. Peyerimhoff, *J. chem. Phys.* **45**, 3682 (1966).
Discussion of Walsh's rules.

HF_2^-

State	Geometry	Basis set	Energy	Wave function	Reference
	R(H–F)				
$^1\Sigma_g^+$	2.153	Min. STO	−198.28264		1
	2.101	GTO	−198.904		
	2.126	Ext. STO	−199.3791	✓	2
	2.1	Ext. STO	−199.5730	✓	5, 6
	2.126	Min. STO + CI	−199.5742		4

Properties	*References*
Orbital energies	1, 2
Nuclear quadrupole coupling	3
Charge density	4
Ionization potential	7

REFERENCES

1. E. Clementi, *J. chem. Phys.* **34**, 1468 (1961).
2. E. Clementi and A.D. McLean, *J. chem. Phys.* **36**, 745 (1962).
3. C.W. Kern and M. Karplus, *J. chem. Phys.* **42**, 1062 (1965).
4. G. Bessis and S. Bratoz, *J. Chim. phys.* **57**, 789 (1960).
5. A.D. McLean and M. Yoshimine, Supp. to *I.B.M. J. Res. Dev.* (1967).
6. M. Yoshimine and A.D. McLean, *Int. J. Quantum Chem.* **1S**, 313 (1967).
7. H. Preuss and R. Janoschek, *J. molec. Struct.* **3**, 423 (1969).

He_3^+

State	Geometry	Basis set	Energy	Wave function	Reference
^2A	R(He–He)				
	2.28	V.B. GTO	−7.083		2

Properties	*References*
Potential curve	1
Spectroscopic constants	2, 3
Binding energy	3

REFERENCES

1. P. Rosen, *J. chem. Phys.* **21**, 1007 (1953).
V.B. treatment for linear and triangular geometry.

2. R.D. Poshusta and D.F. Zetik, *J. chem. Phys.* **48**, 2826 (1968).

3. R.D. Poshusta, J.A. Haugen, and D.F. Zetik, *J. chem. Phys.* **51**, 3343 (1969).
V.B. GTO for Ground state R(He–He) = 2.28.

Li_2O

State	Geometry	Basis set	Energy	Wave function	Reference
	R(Li–O)				
	3.13				
$^1\Sigma$	linear	Ext. STO	–89.7743		1
^1A	$\angle = 150°$		–89.7721		1

Properties	*References*
Charge density	1
Orbital energy	1
Potential curves	1

REFERENCE

1. R.J. Buenker and S.D. Peyerimhoff, *J. chem. Phys.* **45**, 3682 (1966).
Includes discussion of Walsh's rules.

C_3

State	Geometry	Basis set	Energy	Wave function	Reference
	R(C–C)				
$X\,^1\Sigma_g^+$		GTO	–112.49		5
	2.519	Min. STO	–113.0875	✓	1, 3
	2.519	Ext. STO	–113.16518	✓	2

Properties	*References*
Orbital energies	1, 2, 3
Population analysis	3
Term values	4
Ionization potential	5

REFERENCES

1. E. Clementi, *J. chem. Phys.* **34**, 1468 (1961).
2. E. Clementi and A.D. McLean, *J. chem. Phys.* **36**, 45 (1962). Other basis sets included.
3. E. Clementi and H. Clementi, *J. chem. Phys.* **36**, 2824 (1962). Wave functions given in ref. 1.
4. J.W. Nibler and J.W. Linnett, *Trans. Faraday Soc.*, **64**, 1153 (1968). Non-paired spatial orbital method for $X^1\Sigma_g^+$, $^3\Sigma_u^+$, $^1\Sigma_u^+$, $^1\Sigma_u^-$, $^3\Delta_u$, $^1\Delta_u$, and $C_3^+(^2\Pi_u)$, $C_3^-(^2\Pi_g)$.
5. H. Preuss and R. Janoschek, *J. molec. Struct.* **3**, 423 (1969).

HeF_2

Property	*Reference*
Potential curve	1

REFERENCE

1. L.C. Allen, R.M. Erdahl, and J.L. Whitten, *J. Am. chem. Soc.* **87**, 3769 (1965). $^1A_1$ state considered by V.B. with ext. GTO lobe basis.

NeF_2

State	Basis set	Energy	Wave function	Reference
Ground state	V.B.	Repulsive		1

REFERENCE

1. L.C. Allen, A.M. Lesk, and R.M. Erdahl, *J. Am. chem. Soc.* **88**, 615 (1966).

N_3^-

State	Geometry	Basis set	Energy	Wave function	Reference
	R(N–N)				
$X^1\Sigma_g^+$	2.116522	Min. STO	−162.5420		5
	2.217	Min. STO	−162.5422	✓	1
	2.27	Ext. STO	−162.7048	✓	2
	2.20	GTO	−163.1123		4

Properties	*References*
Orbital energies	1, 2, 4, 5
Quadrupole coupling constant	3
Potential surface	4
Dipole moment	5
Charge density	5

REFERENCES

1. E. Clementi, *J. chem. Phys.* **34**, 1468 (1961).
2. E. Clementi and A.D. McLean, *J. chem. Phys.* **39**, 323 (1963).
3. C.W. Kern and M. Karplus, *J. chem. Phys.* **42**, 1062 (1965). Wave functions as in ref. 2.
4. S.D. Peyerimhoff and R.J. Buenker, *J. chem. Phys.* **47**, 1953 (1967). Excited states included.
5. R. Bonaccorsi, C. Petrongolo, E. Scrocco and J. Tomasi, *J. chem. Phys.* **48**, 1500 (1968).

NCO^-

State	Geometry		Basis set	Energy	Wave function	Reference
	R(C–N)	R(C–O)				
$^1\Sigma^+$	2.2110	2.3244	Min. STO	–166.4588		8
	2.213	2.281	Ext. STO	–167.2696		1
	2.213	2.281	Ext. STO	–167.2698	✓	2

Properties	*References*
Dipole moment	1, 3
Orbital energies	2, 3, 5
Atomic population	3
Charge density	3
Quadrupole coupling	3
Electric field gradient	4

REFERENCES

1. M. Yoshimine and A.D. McLean, *Int. J. Quantum Chem.* **1S**, 313 (1967).
2. A.D. McLean and M. Yoshimine. Supp. to *I.B.M. J. Res. Dev.* (1967). Near Hartree-Fock limit.
3. R. Bonaccorsi, C. Petrongolo, E. Scrocco, and J. Tomasi. *J. chem. Phys.* **48**, 1500 (1968). Min STO.
4. R. Bonaccorsi, E. Scrocco, and J. Tomasi, *J. chem. Phys.* **50**, 2940 (1969). Electric field gradient.
5. E. Clementi and D. Klint, *J. chem. Phys.* **50**, 4899 (1969). Uses wave function of ref. 2.

SCN^-

State	Geometry R(C–N)	Geometry R(S–C)	Basis set	Energy	Wave function	Reference
$^1\Sigma^+$	2.30	2.95	Ext. STO	–489.9017	✓	1, 2

Properties	*References*
Dipole moment	1
Orbital energies	2, 4
Electric field gradient	2

REFERENCES

1. M. Yoshimine and A.D. McLean, *Int. J. Quantum Chem.* **1S**, 313 (1967). Near Hartree–Fock wave function.
2. A.D. McLean and M. Yoshimine. Supp. to *I.B.M. J. Res Dev.* (1967). Near Hartree–Fock limit.
3. R. Bonaccorsi, E. Scrocco, and J. Tomasi, *J. chem. Phys.* **50**, 2940 (1969). Electric field gradient with wave function of ref. 2.
4. E. Clementi and D. Klint, *J. chem. Phys.* **50**, 4899 (1969). Uses wave function of ref. 2.

BeF_2

State	Geometry	Basis set	Energy	Wave function	Reference
	R(Be–F)				
$^1\Sigma^+$	2.702	Gaussian lobe	–213.6527		1

Properties	*Reference*
Potential curve	1
Orbital energies	1

REFERENCE

1. S.D. Peyerimhoff, R.J. Buenker, and J.L. Whitten, *J. chem. Phys.* **46**, 4707 (1967).

CO_2

State	Geometry	Basis set	Energy	Wave function	Reference
	R(C–O)				
$^1\Sigma_g^+$	2.1944	Min. STO	–186.8428	✓	1
	2.1944	STO	–187.0757	✓	3
	2.196	Gaussian lobe	–187.4929		5
	2.1944	Ext. STO	–187.7228		6
	2.1944	Ext. STO	–187.7228	✓	7

Properties	*References*
Orbital energies	1, 3, 5, 7
Population analysis	2, 4
Quadrupole coupling	3
Potential curve	5, 7

REFERENCES

1. A.D. McLean, *J. chem. Phys.* **32**, 1595 (1960).
Min. STO.

2. A.D. McLean, B.J. Ransil, and R.S. Mulliken, *J. chem. Phys.* **32**, 1873 (1960).
Population analysis using wave function of ref. 1.

3. A.D. McLean, *J. chem. Phys.* **38**, 1347 (1963).

4. M. Horani, S. Leach, J. Rostas, and G. Berthier, *J. Chim. Phys.* **63**, 1015 (1966).
Uses atomic population to discuss structure of ion.

5. S.D. Peyerimhoff, R.J. Buenker, and J.L. Whitten, *J. chem. Phys.* **46**, 1707 (1967).
Discusses Walsh's rules.

6. M. Yoshimine and A.D. McLean. *Int. J. Quantum Chem.* **1S**, 313 (1967).
Very near Hartree–Fock limit.

7. A.D. McLean and M. Yoshimine, Supp. to *I.B.M. J. Res. Dev.* (1967).
Very near Hartree–Fock limit.

FCN

State	Geometry		Basis set	Energy	Wave function	Reference
	R(F–C)	R(C–N)				
$^1\Sigma^+$	2.3811	2.2016	Ext. STO	−191.7798	✓	1,2

Properties	*Reference*
Dipole moment	1
Orbital energies	2, 4
Electric field gradient	3

REFERENCES

1. M. Yoshimine and A.D. McLean, *Int. J. Quantum Chem.* **1S**, 313 (1967).
Near Hartree–Fock limit.

2. A.D. McLean and M. Yoshimine, Supp. to *I.B.M. J. Res. Dev.* (1967).
Near Hartree–Fock limit.

3. R. Bonaccorsi, E. Scrocco and J. Tomasi, *J. chem. Phys.* **50**, 2940 (1969).
Electric field gradient.

4. E. Clementi and D. Klint, *J. chem. Phys.* **50**, 4899 (1969) .
Discusses orbital energies in various –CN molecules.

N_2O

State	Geometry		Basis set	Energy	Wave function	Reference
	R(N–N)	R(N–O)				
$^1\Sigma_g^+$	2.1273	2.2418	GTO	−183.5763		4
			Ext. STO	−183.7567	✓	2, 3

Properties	*References*
Magnetic properties	1
Dipole moment	2
Orbital energies	3
Potential curve	3, 4
Charge density	4

REFERENCES

1. A.D. McLean and M. Yoshimine, *J. chem. Phys.* **45**, 3676 (1966). Magnetic properties.
2. M. Yoshimine and A.D. McLean, *Int. J. Quantum Chem.* **1S**, 313 (1967).
3. A.D. McLean and M. Yoshimine, Supp. to *I.B.M. J. Res. Dev.* (1967). Very near Hartree–Fock limit.
4. S.D. Peyerimhoff and R.J. Buenker, *J. chem. Phys.* **49**, 2473 (1968). Gaussian basis set.

SCO

State	Geometry		Basis set	Energy	Wave function	Reference
	R(S–C)	R(C–O)				
$X\,^1\Sigma^+$	2.95	2.19	Min. STO	−508.4918	√	1
	2.9442	2.2016	Ext. STO	−510.3309	√	3, 4

Properties	*References*
Dipole moment	3, 5
Orbital energies	1, 4
Excited states	1
Quadrupole coupling	5
Population analysis	2

REFERENCES

1. E. Clementi, *J. chem. Phys.* **36**, 750 (1962). Min. STO
2. M. Horani, J. Rostas, and G. Berthier, *J. Chim. phys.* **63**, 1015 (1966). Discusses structure of ions.
3. M. Yoshimine and A.D. McLean, *Int. J. Quantum Chem.* **1S**, 313 (1967).
4. A.D. McLean and M. Yoshimine, Supp. to *I.B.M. J. Res. Dev.* (1967). Near Hartree–Fock limit.

5. A.D. McLean and M. Yoshimine, *J. chem. Phys.* **46**, 3682 (1967). Uses wave function of ref. 2 and 3.

ClCN

State	Geometry		Basis set	Energy	Wave function	Reference
	R(C–Cl)	R(C–N)				
$^1\Sigma^+$	3.0784	2.1978	Ext. STO	–551.8247	✓	1, 2

Properties	*References*
Dipole moment	1
Orbital energies	2, 4
Electric field gradient	3

REFERENCES

1. M. Yoshimine and A.D. McLean, *Int. J. Quantum Chem.* **1S**, 313 (1967).

2. A.D. McLean and M. Yoshimine, Supp. to *I.B.M. J. Res. Dev.* (1967). Near Hartree–Fock limit.

3. R. Bonaccorsi, E. Scrocco, and J. Tomasi, *J. chem. Phys.* **50**, 2940 (1969). Electric field gradient.

4. E. Clementi and D. Klint. *J. chem. Phys.* **50**, 4899 (1969). Discusses orbital energies in various - CN molecules.

NO_2

State	Geometry		Basis set	Energy	Wave function	Reference
	R(N–O)	∠				
$X\,^2A_1$	2.247	133.6°	GTO	–203.6930		1
	2.249	134°	GTO	–203.729	✓	2
$^2\Sigma_u^+$	2.249		GTO	–203.696		2
$A\,^2B_1$		150°	GTO	–203.679		2
$B\,^2B_2$		90°	GTO	–203.632		2
$^2B_1$		90°	GTO	–203.522		2
$^2\Sigma_g^+$			GTO	–203.478		2
$^2A_1$		90°	GTO	–203.116		2
$^2B_2$		120°	GTO	–202.968		2

Properties	*References*
Orbital energies	1, 2
Force constant	1
Excited states	1, 2
Spectroscopic constants	2

REFERENCES

1. L. Burnelle, A.M. May and R.A. Gangi, *J. chem. Phys.* **49**, 561 (1968). Walsh diagrams and calculations for excited states.
2. W. H. Fink., *J. chem. Phys.* **49**, 5054 (1968). Discussion of excited states.

NO_2^+

State	Geometry	Basis set	Energy	Wave function	Reference
	R(N–O)				
$^1\Sigma_g^+$	2.173	Min. STO	–202.9011	√	1
	2.173	Ext. STO	–203.1082	√	2
	2.1806	Gaussian lobe	–203.4547	√	3

Properties	*References*
Orbital energies	1, 2, 3
Potential curves	3
Charge density	3
Population analysis	3

REFERENCES

1. E. Clementi, *J. Chem. Phys.* **34**, 1468 (1961).
2. E. Clementi and A.D. McLean, *J. chem. Phys.* **39**, 323 (1963). Extended STO
3. G.V. Pfeiffer and L.C. Allen, *J. chem. Phys.* **51**, 190 (1969). Discussed Walsh diagrams.

NO_2^-

State	Geometry		Basis set	Energy	Wave function	Reference
	R(N–O)	∠				
$^1A_1$	2.336	117°	Min. STO	–203.1737		1

State	Geometry		Basis set	Energy	Wave function	Reference
	R(N–O)	∠				
	2.336	115.4°	Gaussian lobe	–203.9551	✓	3
	2.336	117°	STO	–203.9859		2

Properties	*References*
Dipole moment	1, 2
Quadrupole coupling	1, 2
Population analysis	1, 2, 3
Ionization potential	2
Charge density	2, 3
Potential curve	3
Orbital energies	3

REFERENCES

1. R. Bonaccorsi, C. Petrongolo, E. Scrocco, and J. Tomasi, *J. chem. Phys.* **48**, 2107 (1968).
2. R. Bonaccorsi, C. Petrongolo, E. Scrocco, and J. Tomasi, *J. chem. Phys.* **48**, 1497 (1968). Extended STO.
3. G.V. Pfeiffer and L. C. Allen, *J. chem. Phys.* **51**, 190 (1969). Discussed Walsh diagrams.

PO_2^-

State	Geometry		Basis set	Energy	Wave function	Reference
	R(P–O)	∠				
$^1A_1$	2.736	120°	Min. STO	–488.77	✓	1

Properties	*References*
Orbital energies	1
Charge density	1

REFERENCE

1. D.B. Boyd and W.N. Lipscomb, *J. chem. Phys.* **48**, 4968 (1968).

CF_2

State	Geometry		Basis set	Energy	Wave function	Reference
	R(C–F)	∠				
$^1A_1$	2.457	104.8°	GTO	−236.6494		1

Properties	*Reference*
Orbital energies	1
Force constants	1
Walsh diagram	1
Population analysis	1

REFERENCE

1. L.M. Sachs, M. Geller, and J.J. Kaufman, *J. chem. Phys.* **51**, 2771 (1969).

O_3

State	Geometry		Basis set	Energy	Wave function	Reference
	R(O–O)	∠				
$^1A_1$			Min. STO	−223.4790		3
	2.50	116.8°	GTO	−224.1618		2

Properties	*References*
Ionization potential	1
Population analysis	1, 3
Dipole moment	1, 3
Dissociation energy	1
Orbital energies	1, 2, 3
Potential surface	2
Charge densities	3
Quadrupole moment	3

REFERENCES

1. I. Fischer-Hjalmars, *Ark. Fys.* **11**, 529 (1957). Numerical wave function for $^1A_1$ state given. Excited states $^1B_2$, $^1A_1$ and $^1B_1$ treated.

2. S.D. Peyerimhoff and R.J. Buenker, *J. chem. Phys.* **47**, 1953 (1967). Some excited states given.

3. C. Petrongolo, E. Scrocco, and J. Tomasi, *J. chem. Phys.* **48**, 407 (1968).
Coordinates given in paper.

FNO

State	Geometry		Basis set	Energy	Wave function	Reference
	R(F–N) 2.872					
Ground state	R(N–O) 2.135	∠ 110°	Min. STO	–227.7084		2
			Gaussian lobe	–228.3800	√	1

Properties	*References*
Potential surface	1
Spectroscopic constants	1
Charge densities	1, 2
Dipole moment	1, 2
Quadrupole moment	1, 2
Orbital energies	2
Population analysis	2

REFERENCES

1. S.D. Peyerimhoff and R.J. Buenker, *Theor. Chim. Acta* **9**, 103 (1967).
2. C. Petrongolo, E. Scrocco, and J. Tomasi, *J. chem. Phys.* **48**, 407 (1968).
Coordinates given in paper.

SO_2

Properties	*Reference*
Dipole moment	1
Ionization potentials	1

REFERENCE

1. I.H. Hillier and V.R. Saunders. *Chem. Phys. Lett.* **4**, 163 (1969).

F_2O

State	Geometry		Basis set	Energy	Wave function	Reference
	R(F–O)	∠				
$^1A_1$	2.64	102°	Min. STO	–272.4251		2
			STO	–273.4456		1
			Ext. STO	–273.5261		3

Properties	*References*
Dipole moment	2, 3
Orbital energies	1, 3
Charge density	1, 2
Potential curves	1
Quadruple moment	2, 3
Atomic population	2, 3

REFERENCES

1. R.J. Buenker and S.D. Peyerimhoff, *J. chem. Phys.* **45**, 3692 (1966). Discussion of Walsh's rules.
2. C. Petrongolo, E. Scrocco, and J. Tomasi, *J. chem. Phys.* **48**, 407 (1968).
3. R. Bonaccorsi, C. Petrongolo, E. Scrocco, and J. Tomasi, *J. chem. Phys.* **48**, 1497 (1968).

CuF_2

State	Geometry	Basis set	Energy	Wave function	Reference
	R(Cu–F)				
$^2\Sigma^+$	3.324	Min. GTO			1

Properties	*Reference*
Spectroscopic constants	1
Excited states	1

REFERENCE

1. H. Basch, C. Hollister, and J.W. Moskowitz. *Chem. Phys. Lett.* **4**, 79 (1969).
Virtual orbitals used for excited states.

H_4

State	Geometry	Basis set	Energy	Wave function	Reference
$^1\Sigma_g^+$	linear				
	R(H–H)				
	1.6	GTO	–2.1669		3
	1.8	Double zeta	– 2.232		6
	(also other geometries)				

Property	*References*
Potential surface	3, 4, 5, 6

REFERENCES

1. R. Taylor, *Proc. phys. Soc.* **A64**, 249 (1951).
   Simple M.O. calculation.
2. R.S. Barker, R.L. Snow, and H. Eyring, *J. chem. Phys.* **23**, 1686 (1955).
   Comparison of M.O. and V.B. methods.
3. M.E. Schwartz and L.J. Schaad, *J. chem. Phys.* **48**, 4709 (1968).
4. R.N. Porter and L.M. Raffi, *J. chem. Phys.* **50**, 5216 (1969).
   Extension of Woodward Hoffman rules to $H_2 + D_2$.
5. H. Conroy and G. Malli, *J. chem. Phys.* **50**, 5049 (1969).
   Square and rectangular H_4. Correlated wave function.
6. M. Rubinstein and I. Shavitt, *J. chem. Phys.* **51**, 2014 (1969).

H_4^+

REFERENCE

1. M.E. Schwartz and L.J. Schaad, *J. chem. Phys.* **48**, 4709 (1968).
   GTO, $E = -1.7443$.

H_4^{++}

REFERENCE

1. H. Conroy and G. Malli, *J. chem. Phys.* **50**, 5049 (1969).
   Square and rectangular, correlated wave function and energy surface given.

H_4^{+++}

REFERENCE

1. H. Conroy and G. Malli, *J. chem. Phys.* **50**, 5049 (1969).
Square and rectangular ^2A_g states and $^2B_{2u}$, $^2B_{2u}$, and ^2E_u states. Correlated wave function and energy surfaces given.

HeH_3^+

Properties	*References*
Potential surface	1
Spectroscopic constants	2
Binding energy	2

REFERENCES

1. I. Funke, H. Preuss, and G. Diercksen, *Molec. Phys.* **13**, 517 (1967). Hypersurface of reaction

$H_2 + HeH^+ \rightarrow H_3^+ + He$ using GTO.

2. R.D. Poshusta, J.A. Haugen, and D.F. Zetik, *J. chem. Phys.* **51**, 3343 (1969).
V.B. calculation with GTO on $^1A_1$. Geometry given in paper.

LiH_3^+

Properties	*Reference*
Spectroscopic constants	1
Binding energy	1

REFERENCE

1. R.D. Poshusta, J.A. Haugen, and D.F. Zetik, *J. chem. Phys.* **51**, 3343 (1969).
Valence bond with GTO basis for $^2A_1$ state.

BeH_3^-

State	Geometry	Basis set	Energy	Wave function	Reference
	R(Be–H)				
$^1A_1$	2.5	One centre STO	−16.14405	✓	2
	2.54	Ext. GTO	−16.2540		1

Properties	*References*
Potential surface	2
Orbital energies	2
Magnetic properties	4

REFERENCES

1. S.D. Peyerimhoff, R.J. Buenker, and L.C. Allen, *J. chem. Phys.* **45**, 734 (1966).

2. B.D. Joshi, *J. chem. Phys.* **46**, 879 (1967).

3. A.A. Frost, *J. phys. Chem. Ithaca*, **72**, 1289 (1968).
Floating spherical gaussian calculation.

4. D.N. Tripathi, D. Tiwari, and D.K. Rai, *Indian J. pure appl. Phys.* **7**, 707 (1969).
One centre.

BH_3

State	Geometry	Basis set	Energy	Wave function	Reference
	R(B–H)				
$^1A'_1$	2.191	One centre	–26.2358		8
	2.25	Min. STO	–26.3377	✓	6
	2.31	Gaussian lobe	–26.3453		9
	2.26	STO	–26.3517	✓	5
	2.25	Min. STO	–26.3533		11
	2.31	Gaussian lobe	–26.35438		10
	2.31	Gaussian lobe	–26.36520	✓	4
	2.243	Ext. GTO	–26.3734		2
	2.3	Min. STO + CI	–26.38746	✓	3

Properties	*References*
Orbital energies	4, 5, 9
Charge density	5, 6, 9
Magnetic properties	12

REFERENCES

1. W.L. Clinton and B. Rice, *J. chem. Phys.* **29**, 445 (1958).
Semi-empirical.

2. J.J. Kaufman and L.A. Burnelle, *RIAS Tech. Rep.* 65 (1966).
Extended GTO.

3. A. Pipano and I. Shavitt, *Int. J. Quantum Chem.* **2**, 791 (1966).
Includes discussion of convergence of CI.

4. S.D. Peyerimhoff, R.J. Buenker and L.C. Allen, *J. chem. Phys.* **45**, 734 (1966).

5. W.E. Palke and W.N. Lipscomb, *J. chem. Phys.* **45**, 3948 (1966).

6. W.E. Palke and W.N. Lipscomb, *J. Am. chem. Soc.* **88**, 2384 (1966).

7. A.A. Frost, B.H. Prentice, and R.A. Rouse, *J. Am. chem. Soc.* **89**, 3064 (1967).
Floating spherical gaussians.

8. B.D. Joshi, *J. chem. Phys.* **46**, 875 (1967).

9. S.D. Peyerimhoff, R.J. Buenker and J.L. Whitten, *J. chem. Phys.* **46**, 1707 (1967).
Compares results with Walsh's rules.

10. J.D. Petke and J.L. Whitten, *J. chem. Phys.* **51**, 3166 (1969).

11. E. Switkes, R.M. Stevens, and W.N. Lipscomb, *J. chem. Phys.* **51**, 5229 (1969).

12. D.N. Tripathi, D. Tiwari, and D.K. Rai, *Indian J. pure appl. Phys.* **7**, 707 (1969).
One centre.

BH_3^-

State	Geometry	Basis set	Energy	Wave function	Reference
	R(B–H)				
$^2A_1'$	2.5	One centre	–16.144047	✓	2
	2.54	Gaussian lobe	–16.25404	✓	1

Properties	*References*
Orbital energies	2
Hyperfine constants	3

REFERENCES

1. S.D. Peyerimhoff, R.J. Buenker, and L.C. Allen, *J. chem. Phys.* **45**, 734 (1966).

2. B.D. Joshi, *J. chem. Phys.* **46**, 875 (1967).

3. T.A. Claxton, D. McWilliams, and N.A. Smith, *Chem. Phys. Lett.* **4**, 505 (1969).
Unrestricted Hartree–Fock.

CH_3

State	Geometry	Basis set	Energy	Wave function	Reference
	R(C–H)				
$^2A_1'$	2.041	GTO	–39.19335		4
	2.013	STO	–39.3386		2
	2.0126	One centre STO + CI	–39.3715	✓	3
		GTO	–39.5114		6
	2.022	GTO	–39.57148		5

Properties	*References*
Orbital energies	3, 5
Hyperfine splitting constant	3, 6, 7
Potential surface	4
Ionization potential	5

REFERENCES

1. J.R. Hoyland and F.W. Lampe, *J. chem. Phys.* **37**, 1066 (1957). One centre STO, relative energies of CH_3^+, CH_3, CH_4, CH_5^+ given.
2. P.G. Lykos, R.B. Hermann, J.D.S. Ritter, and R. Moccia, *Bull. Am. phys. Soc.* **9**, 145 (1964).
3. A.L.H. Chung, *J. chem. Phys.* **46**, 3144 (1966).
4. K. Morokuma, L. Pedersen, and M. Karplus, *J. chem. Phys.* **48**, 4801 (1968).
5. P. Millie and G. Berthier, *Int. J. Quantum Chem.* **2S**, 67 (1968).
6. T.A. Claxton, D. McWilliams, and N.A. Smith, *Chem. Phys. Lett.* **4**, 505 (1969). Unrestricted Hartree–Fock.
7. W. Mayer, *J. chem. Phys.* **51**, 5749 (1969).

CH_3^+

State	Geometry	Basis set	Energy	Wave function	Reference
	R(C–H)				
$^1A_1$	2.062	Floating spherical gaussian	–33.300		9
	2.013	One centre STO	–39.0189		2

State	Geometry	Basis set	Energy	Wave function	Reference
	R(C–H)				
	2.032	One centre ext. STO	–39.117795		4
	2.040	GTO	–39.144		12
	2.049	GTO	–39.151822		11
	2.1	Gaussian lobe	–39.1646		5
	2.1	Gaussian lobe	–39.1996	✓	3
	1.95	GTO	–39.21726	✓	6
	2.039	GTO	–39.24592		8, 10

Properties	*References*
Potential surface	3, 5, 8, 10
Orbital energies	3, 8, 10
Ionization potential	4, 7, 12
Spectroscopic constants	5, 7, 12
Charge density	6
Population analysis	6
Dipole moment	6

REFERENCES

1. J.R. Hoyland and F.W. Lampe, *J. chem. Phys*. **37**, 1066 (1957). One centre STO. Relative energies on CH_3^+, CH_3, CH_4, CH_5^+ given.

2. P.G. Lykos, R.B. Hermann, J.D.S. Ritter, and R. Moccia, *Bull. Am. phys. Soc*. **9**, 145 (1964).

3. S.D. Peyerimhoff, R.J. Buenker, and L.C. Allen, *J. chem. Phys*. **45**, 734 (1966).

4. B.D. Joshi, *J. chem. Phys*. **46**, 875 (1967).

5. S.D. Peyerimhoff, R.J. Buenker, and J.L. Whitten, *J. chem. Phys*. **46**, 1707 (1967).

6. R.E. Kari and I.G. Csizmadia, *J. chem. Phys*. **46**, 1817 (1967). Pyramidal configuration also calculated. Also some work on excited states.

7. G. Von Bünau, G. Diercksen, and H. Preuss, *Int. J. Quantum Chem*. **1**, 645 (1967). *R*(C–H) = 2.05. GTO basis for $^1A_1$ state.

8. P. Millie and G. Berthier, *Int. J. Quantum Chem*. **2S**, 67 (1968).

9. A.A. Frost, *J. phys. Chem., Ithaca*, **72**, 1289 (1968).

10. R.E. Kari and I.G. Csizmadia, *J. chem. Phys*. **50**, 1443 (1969).

11. J.E. Williams, R. Sustman, L.C. Allen, and P. von R. Schleyer, *J. Am. chem. Soc.* **91**, 1037 (1969).

12. H. Preuss and R. Janoschek, *J. molec. Struct.* **3**, 423 (1969).

CH_3^{++}

Property	*Reference*
Magnetic properties	1

REFERENCE

1. D.N. Tripathi, D. Tiwari, and D.K. Rai, *Indian J. pure appl. Phys.* **7**, 707 (1969).
One centre.

CH_3^-

State	Geometry R(C–H)	Geometry ∠	Basis set	Energy	Wave function	Reference
$^1A_1$	2.283	87.5°	Floating spherical gaussian	–33.315		6
	1.935	90°	STO + CI	–39.228713	√	3
	1.95		GTO	–39.4798		4
	1.95		GTO	–39.50352		7
	1.945	107°	GTO	–39.5125		5
$^1A'_1$	1.935	120°	STO + CI	–39.236059	√	3
	2.013	120°	STO	–39.2734		2

Properties	*References*
Orbital energies	3, 4, 5, 7
Population analysis	4
Dipole moment	4
Charge density	4
Potential surface	4, 5, 7
Spectroscopic constants	7
Magnetic properties	8

REFERENCES

1. H. Hartmann and G. Gliemann, *Z. phys. Chem.* **19**, 29 (1959).
One-centre calculation of energy.

2. P.G. Lykos, R.B. Hermann, J.D.S. Ritter, and R. Moccia, *Bull. Am. phys. Soc.* **9**, 145 (1964).

3. M. Rutledge and F. Saturno, *J. chem. Phys.* **43**, 597 (1965).

4. R.E. Kari and I.G. Csizmadia, *J. chem. Phys.* **46**, 4585 (1967). Inversion barrier calculated. Some excited states also calculated.

5. P. Milli and G. Berthier, *Int. J. Quantum Chem.* **2S**, 67 (1968).

6. A.A. Frost, *J. phys. Chem., Ithaca,* **72**, 1289 (1968).

7. R.E. Kari and I.G. Csizmadia, *J. chem. Phys.* **50**, 1443 (1969). Variation of barrier height with basis set given.

8. D.N. Tripathi, P. Tiwari, and D.K. Rai, *Indian J. pure appl. Phys.* **7**, 707 (1969). One centre.

NH_3

State	Geometry R(N–H)	∠	Basis set	Energy	Wave function	Reference
$^1A_1$	1.904		FSGO	–47.568		25, 32
	Experimental		Min. GTO	–55.0144	√	7
		106° 46′	One centre STO	–55.276		1
	1.9161	106° 46′	One centre STO	–55.6052	√	8
	1.8987		One centre STO	–55.6778	√	12
	1.86	104° 10′	One centre STO	–55.8688	√	14
	1.926		One centre STO	–55.9548		6
	1.928		One centre STO	–55.9748	√	9
	1.93	109° 28′	Min. STO	–56.0033	√	33
			Min. STO	–56.0052	√	16
	1.967	103° 42′	Min. STO	–56.0075	√	13
	1.967		Contr. GTO	–56.0104	√	19
			GTO	–56.03		38
			STO	–56.0301		37
			V.B. STO	–56.0565		15
	1.867	109° 34′	One centre STO	–56.0842	√	10
			Min. STO	–56.096	√	4
			GTO	–56.1415		35

State	Geometry		Basis set	Energy	Wave function	Reference
	R(N–H)	∠				
	1.9162		Gaussian lobe	–56.1460		24
	1.9162		Contr. GTO	–56.1568		30
	1.959	107°	GTO	–56.1812		28
	1.9164	106° 44′	Ext. GTO	–56.183		18
	Experimental		Ext. GTO	–56.201		21
	1.881	114° 6′	Ext. GTO	–56.2109	✓	29
	1.905	109° 3′	Ext. STO	–56.2268	✓	11
	1.916	106° 47′	Min. STO	–56.266	✓	3
			This value may be erroneous			
	Planar					
$^1A_1'$	1.980		FSGO	–47.141		25, 32
	1.8783		One centre STO	–55.6773	✓	12
	1.928		One centre STO	–55.9721	✓	9
	1.967		Min. STO	–55.9891		13
			Contr. GTO	–55.9942		
	1.853		One centre STO	–56.0431	✓	10
	1.916		Min. STO	–56.120		3
	1.9162		Contr. GTO	–56.1541		30
	1.881		Ext. GTO	–56.1993		29

Properties	*References*
X-ray form factor	2
Dipole moment	3, 4, 6, 9, 10, 13, 15, 16, 18, 25, 30, 37
Inversion barrier	3, 9, 10, 12, 13, 22, 24, 29, 30
Orbital energies	3, 4, 19, 32
Magnetic properties	5, 36
Force constant	8, 14, 29
Proton affinity	11, 28
Atomic population	13, 16, 24, 30, 32, 37
Potential curve	15
Charge density	15, 16
Excited states	15
Localized orbitals	17, 26, 35
Quadrupole coupling	20, 27, 34
Polarizability	23, 31

REFERENCES

1. K.E. Banyard and N.H. March, *J. chem. Phys.* **26**, 977 (1956).
One centre STO.

2. K.E. Banyard and N.H. March, *Acta crystallogr.* **9**, 385 (1956).
X-ray form factor using wave function of ref. 1.

3. H. Kaplan, *J. chem. Phys.* **26**, 1704 (1957).
Min. STO: later references suggest these results are incorrect.

4. A.B.F. Duncan, *J. chem. Phys.* **27**, 423 (1957).
Min STO using Slater's rules.

5. K.E. Banyard, *J. chem. Phys.* **33**, 832 (1960).
Diamagnetism.

6. R. Moccia, *J. chem. Phys.* **37**, 910 (1962).
One centre STO.

7. C.M. Reeves and M.C. Harrison, *J. chem. Phys.* **39**, 11 (1963).
Min. GTO.

8. D.M. Bishop, J.R. Hoyland, and R.G. Parr, *Molec. Phys.* **6**, 467 (1963).
One centre min. STO.

9. R. Moccia, *J. chem. Phys.* **40**, 2176 (1964).
One centre STO.

10. B.D. Joshi, *J. chem. Phys.* **43**, S40 (1965).
One centre STO.

11. P. Rajagopal, *Z. Naturf.* **20a**, 1557 (1965).
Near Hartree–Fock limit.

12. R.M. Rutledge and A.F. Saturno, *J. chem. Phys.* **44**, 977 (1966).

13. U. Kaldor and I. Shavitt, *J. chem. Phys.* **45**, 888 (1966).

14. D.M. Bishop, *J. chem. Phys.* **45**, 1787 (1966).

15. H. Morau, *Z. Naturf.* **21a**, 2062 (1966).
Valence bond.

16. W.E. Palke and W.N. Lipscomb, *J. Am. chem. Soc.* **88**, 2384 (1966).
Min. STO from Slater's rules.

17. U. Kaldor, *J. chem. Phys.* **46**, 1981 (1967).
Localized orbitals.

18. L.C. Snyder, *J. chem. Phys.* **46**, 3602 (1967).
Estimates Hartree–Fock limit.

19. E. Clementi, *J. chem. Phys.* **46**, 3851 (1967).
Calculates potential surface for $NH_3 + HCl = NH_4Cl$

20. C.W. Kern, *J. chem. Phys.* **46**, 4543 (1967).
Nuclear quadrupole coupling constant using wave function of ref. 13.

21. C.D. Ritchie and H.F. King, *J. chem. Phys.* **47**, 564 (1967).

22. M.P. Melrose and R.G. Parr, *Theor. Chim. Acta* **8**, 180 (1967) .
Inversion barrier by Hellman–Feynman theorem.

23. G.P. Arrighini, M. Maestro and R. Moccia, *Chem. Phys. Lett.* **1**, 2112 (1967).
Electric polarizability.

24. B. Tinland, *Chem. Phys. Lett.* **2**, 443 (1967).

25. A.A. Frost, B.H. Prentice, and R.A. Rouse, *J. Am. chem. Soc.* **89**, 3064 (1967).
Floating spherical gaussian orbitals.

26. J.G. Stamper and N. Trinajstic, *J. chem. Soc.* 782 (1967).
Localized orbitals.

27. P. Pyykko, *Proc. phys. Soc.* **92**, 841 (1967).
Deutron quadrupole coupling constants.

28. A.C. Hopkinson, N.K. Holbrook, K. Yates, and I.G. Csizmadia, *J. chem. Phys.* **49**, 3596 (1968).
Proton affinity with various basis sets.

29. R.G. Body, D.S. McClure, and E. Clementi, *J. chem. Phys.* **49**, 4916 (1968).

30. A. Veillard, J.M. Lehn, and B. Munsch, *Theor. Chim. Acta* **9**, 275 (1968).
Inversion barrier.

31. G.P. Arrighini, M. Maestro, and R. Moccia., *Symp. Faraday Soc.* **2**, 74 (1968).
Dipole hyperpolarizability.

32. A.A. Frost, *J. phys. Chem., Ithaca,* **72**, 1289 (1968).
Floating spherical gaussian orbitals.

33. P.E. Stevenson and W.N. Lipscomb, *J. chem. Phys.* **50**, 3306 (1969).
Used to calculate wave function for ScH_3NH_3.

34. S. Eleh, T–K Ha, and C.T. O'Konski, *J. chem. Phys.* **51**, 1430 (1969).
Quadrupole coupling.

35. J.P. Petke and J.L. Whitten, *J. chem. Phys.* **51**, 3166 (1969).
Orbital hybridizations.

36. D.N. Tripathi, D. Tiwari, and D.K. Rai, *Indian J. pure appl. Phys.* **7**, 707 (1969).
One centre.

37. E. Switkes, R.M. Stevens, and W.N. Lipscomb, *J. chem. Phys.* **51**, 5229 (1969).
Coordinates given.

38. H. Preuss and R. Janoschek, *J. molec. Struct.* **3**, 423 (1969).

NH_3^+

State	Geometry	Basis set	Energy	Wave function	Reference
	Pyramidal				
$^2A_1$	$R = 1.9087$	STO	–55.7350	√	2
		STO	–55.8348	√	1
	Planar				
	$R = 1.9022$	STO	–55.7326	√	2
		STO	–55.7641		1
		GTO	–55.7991		3

Properties	*Reference*
Charge density	1
Dipole moment	1
Inversion barrier	1, 2
Hyperfine constant	3

REFERENCES

1. J.C. Lorquet and H. Lefebvre-Brion, *J. Chim. phys.* **57**, 85 (1960). Uses atomic basis sets.

2. U. Kaldor and I. Shavitt, *J. chem. Phys.* **45**, 888 (1966).

3. T.A. Claxton, D. McWilliams and N.A. Smith, *Chem. Phys. Lett.* **4**, 505 (1969). Unrestricted Hartree–Fock.

NH_3^{++}

State	Geometry	Basis set	Energy	Wave function	Reference
Ground	Planar	One centre			
state	$R = 2.074$	STO	–54.9808	√	1

Properties	*References*
Orbital energies	1
Potential curves	1
Magnetic properties	2

REFERENCES

1. B.D. Joshi, *J. chem. Phys.* **46**, 875 (1968).

2. D.N. Tripathi, D. Tiwari, and D.K. Rai, *Indian J. pure appl. Phys.* **7**, 707 (1969).
One centre.

PH_3

State	Geometry		Basis set	Energy	Wave function	Reference
	R(P–H)	∠				
$^1A_1$	2.68		Min. GTO	–341.178		5
	2.682	93° 22′	Min. STO	–341.3094		3
	2.672		One centre STO	–341.3960		1, 2
	2.69	93° 50′	GTO	–342.4559		6
	2.672	planar	One centre STO	–341.3556		2

Properties	*Reference*
Dipole moment	1, 2
Orbital energies	3, 5
Charge density	5
Quadrupole coupling	4
Inversion barrier	2, 6

REFERENCES

1. R. Moccia, *J. chem. Phys.* **37**, 910 (1962).

2. R. Moccia, *J. chem. Phys.* **40**, 2176 (1964).
One centre STO for two geometrical configurations.

3. D.B. Boyd and W.N. Lipscomb, *J. chem Phys.* **46**, 910 (1967).
Min. STO.

4. P. Pyykko, *Proc. phys. Soc.* **92**, 841 (1967).
Deutron quadrupole coupling constant.

5. D.B. Cook and P. Palmieri, *Chem. Phys. Lett.* **3**, 219 (1969).
Min. GTO.

6. J.M. Lehn and B. Munsch, *J. Chem. Soc. D.*, 1326 (1969).
Inversion barrier.

H_3O^+

State	Geometry		Basis set	Energy	Wave function	Reference
	R(O–H)	∠				
A	1.7	90° 2′	FSGO	–64.647		6
	1.8	114° 25′	One centre	–75.8743	✓	4
	1.8	120°	V.B.	–76.174		2
	1.8	120°	STO	–76.1820	✓	1
	1.813	117°	GTO	–76.3066		5
	1.8	120°	Ext. GTO	–76.3213		3

Properties	*References*
Dipole moment	1, 2, 3
Population analysis	1, 2
Orbital energies	3
Inversion barrier	4
Force constant	4
Magnetic properties	7

REFERENCES

1. R. Gralin, *Ark. Fys.* **19**, 1417 (1961).
STO.

2. R. Gralin, *Ark. Fys.* **21**, 1 (1966).
Valence bond.

3. J.W. Moskowitz and M.C. Harrison, *J. chem. Phys.* **43**, 3550 (1965).
Various geometries calculated.

4. D.M. Bishop, *J. chem. Phys.* **43**, 4453 (1965).
One centre calculation.

5. A.C. Hopkinson, N.K. Holbrook, K. Yates, and I.G. Csizmadia, *J. chem. Phys.* **49**, 3596 (1968).
Used to calculate proton affinity of water.

6. A.A. Frost, *J. phys. Chem., Ithaca,* **72**, 1289 (1968).
Floating spherical gaussians.

7. D.N. Tripathi, D. Tiwari, and D.K. Rai, *Indian J. pure appl. Phys.* **7**, 707 (1969).
One centre.

H_3O^{+++}

REFERENCE

1. B.D. Joshi, *J. chem. Phys.* **46**, 875 (1967).
One-centre calculation indicates instability.

H_3S^+

State	Geometry		Basis set	Energy	Wave function	Reference
	R(H–S)	∠				
$^1A_1$	2.523	117°	GTO	−381.5551		1

Property	*Reference*
Proton affinity	1

REFERENCE

1. A.C. Hopkinson, N.K. Holbrook, K. Yates, and I.G. Csizmadia, *J. chem. Phys.* **49**, 3596 (1968).

ScH_3

State	Geometry		Basis set	Energy	Wave function	Reference
Ground	R(Sc–H)	∠				
state	3.294	109° 28′	Min. STO	−760.1079	√	1

Properties	*Reference*
Orbital energy	1
Population analysis	1

REFERENCE

1. P.E. Stevenson and W.N. Lipscomb, *J. chem. Phys.* **50**, 3306 (1969).

TiH_3^+

State	Geometry		Basis set	Energy	Wave function	Reference
	R(Ti–H)	∠				
	3.077	109° 28′	Min. STO	−848.2043	√	1

Properties	*Reference*
Orbital energies	1
Population analysis	1

REFERENCE

1. P.E. Stevenson and W.N. Lipscomb, *J. chem. Phys.* **50**, 3306 (1969).

C_2H_2

State	Geometry		Basis set	Energy	Wave function	Reference
	R(C–C)	R(C–H)				
$^1\Sigma_g^+$	2.002	2.281	Min. STO	–76.5438		1
			Group orbitals	–76.5733		14
	2.05	2.29	GTO	–76.577		19, 20
	2.002	2.281	Min. STO	–76.6165	✓	7
	2.002	2.281	Min. STO	–76.6416		18
			STO	–76.678		13
	2.002	2.281	GTO	–76.74182		5, 6
	2.002	2.272	Gaussian lobe	–76.7916		12
	2.002	2.281	Ext. STO	–76.8450		10
$^3E_{2u}$			STO	–76.419		13

Properties	*References*
Charge density	2, 3, 5, 6, 13
Nuclear quadrupole coupling	4
Orbital energies	5, 10, 19
Nuclear spin–spin coupling	8, 17
Spectroscopic constants	19

REFERENCES

1. A.D. McLean, *J. chem. Phys.* **32**, 1595 (1960).

2. A.D. McLean, B.J. Ransil, and R.S. Mulliken, *J. chem. Phys.* **32**, 1873 (1960).

3. E. Clementi and H. Clementi, *J. chem. Phys.* **36**, 2824 (1962).

4. C.W. Kern and M. Karplus, *J. chem. Phys.* **42**, 1062 (1965).

5. J.W. Moskowitz, *J. chem. Phys.* **43**, 60 (1965).

6. J.W. Moskowitz, *J. chem. Phys.* **45**, 2338 (1961). Errata to 5.

7. W.E. Palke and W.N. Lipscomb, *J. Am. chem. Soc.* **88**, 2384 (1966).

8. C. Barbier and G. Bertier, *Int. J. Quantum Chem.* **1**, 657 (1967).

9. J. Cizek and A. Pellegatti, *Int. J. Quantum Chem.* **1**, 653 (1967). Calculation of correlation energy.

10. A.D. McLean and M. Yoshimine, Supp. to *I.B.M. J. Res. Dev.* (1967).

11. A.A. Frost, B.A. Prentice, and R.A. Rouse, *J. Am. chem. Soc.* **89**, 3064 (1967). Floating spherical gaussian calculation.

12. R.J. Buenker, S.D. Peyerimhoff, and J.L. Whitten, *J. chem. Phys.* **46**, 2029 (1967).

13. M.G. Griffith and L. Goodman, *J. chem. Phys.* **47**, 4494 (1967). Includes calculation on ions.

14. M. Klessinger, *Symp. Faraday Soc.* **2**, 73 (1968).

15. A.A. Frost and R.A. Rouse, *J. Am. chem. Soc.* **90**, 1965 (1968). Floating spherical gaussian calculation.

16. J.R. Hoyland, *J. chem. Phys.* **48**, 5736 (1968). Two centre calculation.

17. T. Yonezawa, I. Morishima, M. Fujii, and H. Kato, *Bull. chem. Soc. Japan* **42**, 1248 (1969). C^{13}–H spin coupling constant.

18. E. Switkes, R.M. Stevens, and W.N. Lipscomb, *J. chem. Phys.* **51**, 5229 (1969).

19. J. Haase, R. Janoschek, H. Preuss and G. Diercksen, *J. molec. Struct.* **3**, 165 (1969).

20. H. Preuss and R. Janoschek, *J. molec. Struct.* **3**, 423 (1969).

HCHO

State	Geometry			Basis set	Energy	Wave function	Reference
	R(C–H)	R(C–O)	$\hat{HCH}$				
$^1A_1$	2.0	2.3	120°	Min. STO	–113.4496	√	4
				STO	–113.5228		17
	2.12	2.29	118°	GTO	–113.5874		12
	2.12	2.29	118°	Min. STO	–113.591	√	2
				GTO + CI	–113.8094		18
	2.12	2.29	118°	GTO	–113.8334	√	11
	2.109	2.284	116°	GTO	–113.8917	√	14

Properties	*References*
Orbital energies	1, 2, 4, 7, 14, 18
Dipole moment	1, 2, 4, 8, 11, 13, 14, 17
Population analysis	2, 11, 14, 17
Electric properties	3, 6, 13, 14
Magnetic properties	5, 6, 13, 14
Quadrupole coupling constant	6
Term values	10
Spectroscopic constants	10
Ionization potential	14
Dissociation energy	14

REFERENCES

1. J.M. Foster and S.F. Boys, *Rev. mod. Phys.* **32**, 303 (1960). Min. STO basis. Wave functions given. Geometry as in ref 4.
2. P.L. Goodfriend, F.W. Birso, and A.B.F. Duncan, *Rev. mod. Phys.* **32**, 307 (1960).
3. J.T. Lowe and W.H. Flygare, *J. chem. Phys.* **41**, 2153 (1964). Wave functions as in ref. 1.
4. M.D. Newton and W.E. Palke, *J. chem. Phys.* **45**, 2329 (1966). Wave functions agree with ref. 1 but not ref. 2.
5. W.H. Flygare and V.W. Weiss, *J. chem. Phys.* **45**, 2785 (1966). Constructed set of hybrid orbitals used–STO basis.
6. W.H. Flygare, J.M. Pochan, G.I. Kerley, T. Caves, M. Karplus, S. Aung, R.M. Pitzer, and S.I. Chan, *J. chem. Phys.* **45**, 2793 (1966). Many one-electron properties from wave functions in ref. 1.
7. S. Aung, R.M. Pitzer, and S.I. Chan, *J. chem. Phys.* **45**, 3457 (1966). Many one-electron properties from wave functions in ref. 1. Refinement of ref. 6.
8. J.G. Stamper and N. Trinajstic, *J. chem. Soc.* (*A*), 782 (1967). Localized orbitals from wave function in ref. 1.
9. W. Hüttner, M-K Lo, and W.H. Flygare, *J. chem. Phys.* **48**, 1206 (1968). One-electron properties from wave functions in ref. 1 and 5.
10. T.H. Dunning and V. McKay, *J. chem. Phys.* **48**, 5263 (1968). States $^1A_1$, $^3A_2$, $^1A_2$, $^3B_1$, $^1B_1$, $^3A_1$, $^1B_2$, $^3B_2$. Min. STO basis. One-electron properties calculated.
11. N.W. Winter, T.H. Dunning and J.H. Letcher, *J. chem. Phys.* **49**, 1871 (1968). Near Hartree–Fock basis functions. Coordinates given.

12. A.C. Hopkinson, N.K. Holbrook, K. Yates, and I.C. Csizmadia, *J. chem. Phys.* **49**, 3596 (1968).
Proton affinity calculated.

13. T.H. Dunning, N.W. Winter, V. McKay, *J. chem. Phys.* **49**, 4128 (1968).
GTO basis.

14. D.B. Newman and J.W. Moskowitz, *J. chem. Phys.* **50**, 2216 (1969).

15. B. Levy, *Chem. Phys. Lett.* **4**, 17 (1969).
Multi-configurational SCF using STO basis (ref. 4).
Wave functions given. Electronic energy = – 145.0040.

16. P. Ros, *J. chem. Phys.* **49**, 4902 (1968).
Proton affinity.

17. E. Switkes, R.M. Stevens, and W.N. Lipscomb, *J. chem. Phys.* **51**, 5229 (1969).
Geometry in paper.

18. J.L. Whitten and M. Hackmeyer, *J. chem. Phys.* **51**, 5584 (1969).
CI on excited states, $^2A_2$, $^{1,3}A_1$, $^1B_1$, $^1B_2$.

H_2NO

Property	*Reference*
Reaction surface	1

REFERENCE

1. M. Krauss, *J. Res. natn. Bur. Stand.* **73A**, 191 (1969).
Interaction between NO and H_2. GTO basis.

H_2O_2

State			Basis set	Energy	Wave function	Reference
^1A	R(O–O)	R(O–H)				
	3.731	1.795				
	$\angle = 120°$		Min. GTO	–149.2882		4
			Min. STO	–150.1565	✓	1
			Min. STO	–150.2232	✓	5
			Gaussian lobe	–150.7078	✓	2
			Gaussian lobe	–150.7375	✓	3
			GTO	–150.7992		8

Properties	*References*
Rotation barrier	1, 2, 3, 4, 5, 7, 8
Dipole moment	1
Orbital energies	1
Atomic population	1, 4
Polarizability	6

REFERENCES

1. U. Kaldor and I. Shavitt, *J. chem. Phys.* **44**, 1823 (1966). Min. STO.
2. W. H. Fink and L.C. Allen, *J. chem. Phys.* **46**, 2261 (1967).
3. W.H. Fink and L.C. Allen, *J. chem. Phys.* **46**, 2276 (1967). Discusses rotation barrier.
4. L. Pedersen and K. Morokuma, *J. chem. Phys.* **46**, 3941 (1967). Min. GTO.
5. W.E. Palke and R.H. Pitzer, *J. chem. Phys.* **46**, 3948 (1967). Calculation of rotation barrier.
6. G.P. Arrighini, M. Maestro, and R. Moccia, *Chem. Phys. Lett.* **1**, 242 (1967).
   Electronic polarizability.
7. M.D. Melrose and R.G. Parr, *Theor. Chim. Acta.* **8**, 150 (1967). Rotation barrier obtained by He llman–Feyman theorem.
8. A. Veillard, *Chem. Phys. Lett.* **4**, 51 (1969). Most accurate calculation of rotation barrier.

H_2SO

State	Geometry		Basis set	Energy	Wave function	Reference
$^1A'$	R(S–O)	2.779				
	R(S–H)	2.5167				
	$\hat{HSH}$	100°				
	$\hat{HSO}$	107°	Ext. GTO	–451.11639		1

Property	*Reference*
Dipole moment	1

REFERENCE

1. A. Ruak and I.G. Csizmadia, *Can. J. Chem.* **46**, 1205 (1968).

H_2S_2

State	Geometry		Basis set	Energy	Wave function	Reference
^1A	R(S–S)	3.88				
	R(S–H)	2.51				
	$\hat{SSH}$	91° 20′				
	Dihedral ∠	90° 45′	GTO	–796.18403		1
	Experimental Dihedral ∠	90°	GTO	–792.6581		2

Properties	*References*
Rotation barrier	1, 2
Population analysis	2

REFERENCES

1. A. Veillard and J. Demuynck, *Chem. Phys. Lett.* **4**, 476 (1969).
2. M.E. Schweitz, *J. chem. Phys.* **51**, 4182 (1969).

LiCCH

State	Geometry		Basis set	Energy	Wave function	Reference
$^1\Sigma^+$	R(C–C)	2.2696				
	R(C–H)	2.0088				
	R(Li–C)	3.55	Ext. STO	–83.73055	✓	1, 2

Properties	*References*
Orbital energies	1, 2
Population analysis	2
Ionization potentials	2
Dipole moment	2
Quadrupole moment	2
Magnetic properties	2

REFERENCES

1. A.D. McLean and M. Yoshimine, Supp. to *I.B.M. Res. Dev.* November (1967).
Hartree–Fock limit.

2. A. Veillard, *J. chem. Phys.* **48**, 2012 (1968).
Hartree–Fock limit.

HN_3

State	Geometry	Basis set	Energy	Wave function	Reference
$^1A'$	R(N–N) 2.1411 2.3376 R(N–H) 1.814	Min. STO	–163.2239		1

Properties	*Reference*
Orbital energy	1
Dipole moment	1
Quadrupole coupling	1
Localized orbitals	1
Spectroscopic constants	1
Charge density	1

REFERENCE

1. R. Bonaccorsi, C. Petrongolo, E. Scrocco, and J. Tomasi, *J. chem. Phys.* **48**, 1500 (1968).

FCCH

State	Geometry	Basis set	Energy	Wave function	Reference
$^1\Sigma^+$	R(F–C) 2.41 R(C–C) 2.26 R(C–H) 2.05	GTO	–174.310		3, 4
	R(F–C) 2.417 R(C–C) 2.264 R(C–H)	Ext. STO	–175.7236	✓	1, 2

Properties	*References*
Dipole moment	2
Orbital energies	1, 3
Spectroscopic constants	3

REFERENCES

1. M. Yoshimine and A.D. McLean, *Int. J. Quantum Chem.* **1S**, 313 (1967).

2. A.D. McLean and M. Yoshimine, Supp. to *I.B.M. J. Res. Dev.* (1967).
Near Hartree–Fock limit.

3. J. Haase, R. Janoschek, H. Preuss, and G. Diercksen, *J. molec. Struct.* **3**, 165 (1969).

4. H. Preuss and R. Janoschek, *J. molec. Struct.* **3**, 423 (1969).

ClCCH

State	Geometry	Basis set	Energy	Wave function	Reference
$^1\Sigma^+$	R(C–Cl) 3.18 R(C–C) 2.23 R(C–H) 2.05	GTO	–528.267		3, 4
	R(C–Cl) 3.084 R(C–C) 2.2885 R(C–H) 1.9880	Ext. STO	–535.7673	✓	1, 2

Properties	*References*
Orbital energy	2, 3
Dipole moment	1
Spectroscopic constants	3, 4

REFERENCES

1. M. Yoshimine and A.D. McLean, *Int. J. Quantum Chem.* **1S**, 313 (1967

2. A.D. McLean and M. Yoshimine, Supp. to *I.B.M. J. Res. Dev.* (1967).
Near Hartree–Fock limit.

3. J. Haase, R. Janoschek, H. Preuss, and G. Diecksen, *J. molec Struct.* **3**, 165 (1969).

4. H. Preuss and R. Janoschek, *J. molec. Struct.* **3**, 423 (1969).

HNCO

State	Geometry	Basis set	Energy	Wave function	Reference
$^1A'$	R(C–O) 2.219 R(C–N) 2.280 R(C–H) 1.848 $H\hat{N}C$ 128°	Min. STO	–167.0757		1

Properties	*Reference*
Orbital energies	1
Dipole moment	1
Charge density	1

REFERENCE

1. R. Bonaccorsi, C. Petrongolo, E, Scrocco, and J. Tomasi, *J. chem. Phys*. **48**, 1500 (1968).
Coordinates given.

HCOF

State	Geometry	Basis set	Energy	Wave function	Reference
$^1A'$	R(C–O) 2.100 R(C–F) 2.126 R(C–H) 1.753 $\hat{OCH}$ 105°	GTO	–212.1139	✓	1

Properties	*Reference*
Orbital energies	1
Dipole moment	1
Potential curves	1
Spectroscopic constants	1
Charge density	1

REFERENCE

1. I.G. Csizmadia, M.C. Harrison, and R.T. Sutcliffe, *Theor. Chim. Acta* **6**, 217 (1966).

C_4

State	Geometry	Basis set	Energy	Wave function	Reference
	R(C–C)				
	2.419	STO	–150.2899		2
$X^3\Sigma_g^-$	2.419	STO	–150.84363	✓	1
$^1\Sigma_g^+$	2.419	STO	–149.62863	✓	1

Properties	*References*
Orbital energies	1, 2
Term values	1
Population analysis	2

REFERENCES

1. E. Clementi and K.S. Pitzer, *J. Am. chem. Soc.* **83**, 4561 (1961). Some treatment of $^1\Delta_g$, $^3\Pi_u$, and $^1\Pi_u$ included.

2. E. Clementi and H. Clementi, *J. chem. Phys.* **36**, 2824 (1962). Wave functions as in ref. 1.

3. H. Preuss and R. Janoschek, *J. molec. Struct.* **3**, 423 (1969). Preliminary energy = –137.87.

NCCN

State	Geometry		Basis set	Energy	Wave function	Reference
	R(N–C)	R(C–C)				
$^1\Sigma_g^+$	2.1867	2.6082	Min. STO	–183.9817	√	1
	2.186	2.608	Ext. STO	–184.6568	√	4

Properties	*References*
Orbital energies	1, 4, 5
Population analysis	2
Charge density	3, 5
Electrical properties	6

REFERENCES

1. E. Clementi and A.D. McLean, *J. chem. Phys.* **36**, 563 (1962). Min. STO.

2. E. Clementi and H. Clementi, *J. chem. Phys.* **36**, 2824 (1962). Population analysis using wave function of ref. 1.

3. L. Burnelle, *Theor. Chim. Acta* **2**, 177 (1964). Charge density using wave function of ref. 1.

4. A.D. McLean and M. Yoshimine, Supp. to *I.B.M. J. Res. Dev.* (1967). Very near Hartree–Fock limit.

5. E. Clementi and D. Klint, *J. chem. Phys.* **50**, 4899 (1969). Comparison with other molecules containing CN group.

6. R. Bonaccorsi, E. Scrocco, and S. Tomasi, *J. chem. Phys.* **50**, 2940 (1969).

BF_3

State	Geometry	Basis set	Energy	Wave function	Reference
	R(B–F)				
$^1A'$	2.36	GTO	–320.312		2
	2.48	Ext. GTO	–322.452087		1

Properties	*References*
Charge densities	1
Ionization potential	2

REFERENCES

1. D.R. Armstrong and P.G. Perkins, *J. chem. Soc. D,* 856 (1969).
2. H. Preuss and R. Janoschek, *J. molec. Struct.* **3**, 423 (1969).

NF_3

State	Geometry	Basis set	Energy	Wave function	Reference
$^1A_1$	Exp.	GTO	–352.01749	✓	1

Properties	*Reference*
Orbital energies	1
Dipole moment	1
Magnetic properties	1
Population analysis	1
Quadrupole moment	1

REFERENCES

1. M.L. Unland, T.H. Dunning, and J.R. von Wazer, *J. chem. Phys.* **50**, 3208, 3214 (1969).
Cartesian coordinates given.

BH_4^-

State	Geometry	Basis set	Energy	Wave function	Reference
	R(B–H)				
$^1A_1$	2.372	Min. STO	–26.9232	✓	4
	2.3	Ext. GTO	–26.9449	✓	2

Properties	*References*
Orbital energies	4
Charge densities	4
Force constants	1
Deutron quadrupole coupling	3
Electric field gradient	3

REFERENCES

1. D.M. Bishop, *Theor. Chim. Acta* **1**, 410 (1963).
One-centre calculation.
2. M. Krauss, *J. Res. natn. Bur. Stand.* **68A**, 635 (1964).
3. P. Pyykko, *Proc. phys. Soc.* **92**, 841 (1967).
BD_4^- using one-centre function.
4. R.A. Hegstrom, W.E. Palke, and W.N. Lipscomb, *J. chem. Phys.* **46**, 920 (1967).
5. A.A. Frost, *J. phys. Chem. Ithaca* **72**, 1289 (1968).
Floating spherical gaussian calculation.

B_2H_6

State	Geometry	Basis set	Energy	Wave function	Reference
^1A	Experimental (Coords given)	Min. STO	−52.6782	√	2
	"	GTO lobe	−52.6896		3
	"	Optimized STO	−52.7151	√ √	4
	"	Optimized STO	−52.7204	√	5

Properties	*References*
Orbital energies	3, 4
Charge density	2, 3, 4, 5
Chemical shift	1

REFERENCES

1. C.W. Kern and W.N. Lipscomb, *J. chem. Phys.* **37**, 275 (1962).
Chemical shift.
2. W.E. Palke and W.N. Lipscomb, *J. Am. chem. Soc.* **88**, 2384 (1966).
3. R.J. Buenker, S.D. Peyerimhoff, L.C. Allen and J.L. Whitten, *J. chem. Phys.* **45**, 2835 (1966).

4. W.E. Palke and W.N. Pipscomb, *J. chem. Phys.* **45**, 3948 (1966).

5. E. Switkes, R.M. Stevens, W.N. Lipscomb, and M.D. Newton, *J. chem. Phys.* **51**, 2085 (1969).

B_4H_4

State	Geometry		Basis set	Energy	Wave function	Reference
Ground state	Tetrahedral					
	R(B–B)	R(B–H)				
	3.288	2.267	Min. STO	–167.3206	✓	1

Properties	*Reference*
Orbital energies	1
Charge density	1

REFERENCE

1. W.E. Palke and W.N. Lipscombe, *J. chem. Phys.* **45**, 3945 (1966).

BH_3NH_3

State	Geometry	Basis set	Energy	Wave function	Reference
	Coordinates given	GTO lobe	–82.517		4
	" "	Ext. GTO	– 82.529		1
	" "	Ext. GTO	– 82.5754	✓	3
	" "	Ext. GTO	–82.59651		2

Properties	*References*
Orbital energies	3, 4
Charge density	2, 4
Barrier to internal rotation	5
Dipole moment	3
Quadrupole tensor	3

REFERENCES

1. A. Veillard, B. Levy, R. Daudel, and F. Gallais, *Theor. Chim. Acta,* **8**, 312 (1967).

2. M-Cl Moireau and A. Veillard, *Theor. Chim. Acta,* **11**, 344 (1968).

3. S.D. Peyerimhoff and R.J. Buenker, *J. chem. Phys.* **49**, 312 (1968). Includes some CI.

4. B. Tinland, *J. molec. Struct.* **3**, 244 (1969).

5. A. Veillard, *Chem. Phys. Lett.* **3**, 128 (1969).

BH_2NH_2

State	Geometry	Basis set	Energy	Wave	Reference
	R(B–H) 2.248 R(N–H) 1.91 R(B–N) 2.56 (Angles 120°)	Min. GTO	–81.4123		1

Properties	*Reference*
Dipole moment	1
Orbital energies	1
Population analysis	1
Potential curve	1

REFERENCE

1. D.R. Armstrong, B.J. Duke, and P.G. Perkins, *J. chem. Soc. A*, 2566 (1969).
Comparison with CNDO etc.

Borazine $B_3N_3H_3$

State	Geometry	Basis set	Energy	Wave function	Reference
$^1A_1'$	D_{3h}	GTO	–240.5260		1
$^1A_1$	C_{2v}	GTO	–240.4515		

Properties	*Reference*
Population analysis	1
Ionization potential	1

REFERENCE

1. D.R. Armstrong and D.T. Clark, *J. chem. Soc. D.*, 99 (1970).

BeB_2H_8

State	Geometry	Basis set	Energy	Wave function	Reference
Ground state	Geometry given	GTO	−1792.128		1

Properties	*Reference*
Atomic population	1
Dipole moment	1

REFERENCES

1. D.R. Armstrong and P.G. Perkins, *J. chem. Soc. D.*, 352 (1968).
2. G. Gundersen and A. Haaland, *Acta chem. scand.* **22**, 867 (1968).

AlH_4^-

State	Geometry	Basis set	Energy	Wave function	Reference
$^1A_1$	*R*(Al–H)				
	3.132	One-centre STO	−242.027		2
	2.965	Numerical one-centre	−243.734		1

Properties	*References*
Force constant	1
Magnetic susceptibility	1
Quadrupole coupling	3

REFERENCES

1. E.L. Albasiny and J.R.A. Cooper, *Proc. phys. Soc.* **85**, 1133 (1965). Numerical one-centre calculation
2. A. Sutton and K.E. Banyard, *Molec. Phys.* **12**, 377 (1967). One-centre STO.
3. P. Pyykko, *Proc. phys. Soc.* **92**, 841 (1969).

SiH_4

State	Geometry	Basis set	Energy	Wave function	Reference
	R(Si–H)				
$^1A_1$	2.52	Min. GTO	–288.971		6
	3.49	One-centre STO	–290.02		1
	2.787	One-centre STO	–290.0826		3
	2.787	One-centre STO	–290.1024	✓	4
	2.795	Min. STO	–290.5187	✓	7
	2.78	Numerical one-centre	–290.8	✓	2

Properties	*References*
Spectroscopic constants	2
Charge density	2
Magnetic properties	2
Dipole moment	3
Quadrupole coupling	5
Orbital energies	6, 7
Charge density	6

REFERENCES

1. C.A. Carter, *Proc. R. Soc.* **A235**, 321 (1956).
   One-centre central field
2. E.L. Albasiny and J.R. Cooper, *Proc. phys. Soc.* **85**, 1133 (1965).
   Numerical one-centre.
3. R. Moccia, *J. chem. Phys.* **37**, 910 (1962).
4. R. Moccia, *J. chem. Phys.* **40**, 2164 (1964).
   One-centre min. STO.
5. P. Pyykko, *Proc. phys. Soc.* **92**, 841 (1967).
   Deuteron quadrupole coupling constant.
6. D.B. Cook and P. Palmieri, *Chem. Phys. Lett.* **3**, 219 (1969).
   Min. STO.
7. F.B. Boer and W.N. Lipscomb, *J. chem. Phys.* **50**, 989 (1969).
   Min. STO.

CH_3SiH_3

State	Geometry	Basis set	Energy	Wave function	Reference
Ground state	Staggered	GTO	−330.2323		1

Property	*Reference*
Rotation barrier	1

REFERENCE

1. A. Veillard, *Chem. Phys. Lett.* **3**, 128 (1969).

NH_4^+

State	Geometry	Basis set	Energy	Wave function	Reference
$^1A_1$	R(N–H)				
	1.876	FSGO	−47.893		1
	1.836	One-centre STO	−55.684		1
	1.835	Min. STO	−55.684		2
	1.90	One-centre STO	−55.967	✓	4
	1.99	One-centre STO	−56.2177	✓	5
		GTO	−56.3529	✓	8
	1.9464	GTO	−56.5038	✓	3
	1.964	GTO	−56.5038	✓	6
	1.905	STO	−56.5216		7
	1.947	GTO	−56.5320		10

Properties	*References*
Force constant	2, 4
Ionization potential	2
Magnetic properties	2
Orbital energies	3, 5, 6, 8
Potential curve	4
Charge density	8
Quadrupole coupling	9

REFERENCES

1. M.J. Bernal, *Proc. phys. Soc.* **66A**, 514 (1953).
Single centre.

2. F. Grein, *Theor. Chim. Acta* **1**, 52 (1962).
Min. STO

3. M. Krauss, *J. chem. Phys.* **38**, 564 (1963).

4. D.M. Bishop, *Theor. Chim. Acta* **1**, 410 (1963).
Potential curve calculated.

5. R. Moccia, *J. chem. Phys.* **40**, 2176 (1964).
One centre STO.

6. M. Krauss, *J. Res. natn. Bur. Stand.* **68A**, 635 (1964).
GTO.

7. P. Rajagopal, *Z. Naturf.* **20a**, 1557 (1965).
Proton affinity of NH_3 calculated.

8. E. Clementi, *J. chem. Phys.* **46**, 3851 (1967).
Contracted GTO.

9. P. Pyykko, *Proc. phys. Soc.* **92**, 841 (1967).
Deutron quadrupole coupling constant.

10. A.C. Hopkinson, N.K. Holbrook, K. Yates, and I.G. Csizmadia, *J. chem. Phys.* **49**, 3596 (1968).
Proton affinity of NH_3 calculated.

11. A.A. Frost, *J. phys. Chem., Ithaca*, **72**, 1289 (1968).
Floating spherical gaussian orbitals.

NH_4Cl

State	Geometry	Basis set	Energy	Wave function	Reference
Ground state		GTO	–515.8314	✓	1

Properties	*References*
Charge density	1
Orbital energies	1
Potential surface	1, 3
Atomic population	2
Dissociation energy	1, 3
Vibration frequencies	3

REFERENCES

1. E. Clementi, *J. chem. Phys.* **46**, 3851 (1967).
Study of reaction $NH_3 + HCl = NH_4Cl$.

2. E. Clementi, *J. chem. Phys.* **47**, 2323 (1967).
Atomic population for the reaction.

3. E. Clementi and J.N. Gayles *J. chem. Phys.* **47**, 3837 (1967).
Uses wave function of ref. 1.

N_2H_4

State	Geometry	Basis set	Energy	Wave function	Reference
$^1A_1$	Exp.	GTO	–111.030		1
		GTO	–111.0743	✓	2
		GTO	–111.1239		3

Properties	*References*
Potential curve	1, 3
Rotational barrier	1, 2, 3
Ionization potential	1
Atomic population	2
Quadrupole coupling	4

REFERENCES

1. A. Veillard, *Theor. Chim. Acta,* **5**, 413 (1963).
2. L. Pedersen and K. Morokuma, *J. chem. Phys.* **46**, 3941 (1967).
3. W.A. Fink, D.C. Pan and L.C. Allen, *J. chem. Phys.* **47**, 895 (1967).
4. C.T. O'Konski and T-K Ha, *J. chem. Phys.* **49**, 5354 (1968).
Uses Gaussian lobe basis set.

NH_2OH

State	Geometry	Basis set	Energy	Wave function	Reference
$^1A'$	Exp.	GTO	–129.7062		1
		GTO	–130.8975	✓	2

Properties	*References*
Rotational barrier	1, 2
Potential curve	2
Atomic population	1
Quadrupole coupling	3

REFERENCES

1. L. Pedersen and K. Morokuma, *J. chem. Phys.* **46**, 3941 (1967).
2. W.H. Fink, D.C. Pan, and L.C. Allen, *J. chem. Phys.* **47**, 895 (1967).
3. C.T. O'Konski and T-K Ha, *J. chem. Phys.* **49**, 5354 (1968).

PH_4^+

State	Geometry	Basis set	Energy	Wave function	Reference
	R(P–H)				
$^1A_1$	2.750	One-centre STO	–341.5493	√	1

Property	*Reference*
Quadrupole coupling	2

REFERENCES

1. R. Moccia, *J. chem. Phys.* **40**, 2176 (1964). One centre STO.
2. P. Pyykko, *Proc. phys. Soc.* **92**, 841 (1967). Deutron quadrupole coupling constant.

H_4O^{++}

REFERENCE

1. A.A. Frost, *J. phys. Chem., Ithaca,* **72**, 1289 (1968). Floating spherical gaussians.

$(H_2O)_2$

State	Geometry	Basis set	Energy	Wave function	Reference
Ground state	Linear hydrogen bond	GTO	–151.1313		1
		GTO	–152.0145		2
		GTO	–152.1117		3

Properties	*Reference*
Potential curve	1
Force constant	1
Dipole moment	1
Population analyses	1

REFERENCES

1. K. Morokuma and L. Pedersen, *J. chem. Phys.* **48**, 3275 (1968). Various geometries calculated.
2. P.A. Kollman and L.C. Allen, *J. chem. Phys.* **51**, 3286 (1969). Discussion of hydrogen bonding.
3. G.H.F. Diercksen, *Chem. Phys. Lett.* **4**, 373 (1969).

$(H_2O)_n$

REFERENCE

1. J. del Bene and J.A. Pople, *Chem. Phys. Lett.* **4**, 426 (1969). STO basis. Intermolecular energies and configuration for small groups of H_2O molecules.

$F(H_2O)_2^-$

REFERENCE

1. H. Preuss and R. Janoschek, *J. molec. Struct.* **3**, 423 (1969). Min. GTO for $n = 1, 2, 3$, and 4.

ScH_3NH_3

State	Geometry	Basis set	Energy	Wave function	Reference
	R(Sc–H) 3.2921 R(Sc–N) 4.252 R(N–H) 1.905 ∠ 109° 28′ staggered	Min. STO	−816.2044	√	1

Properties	*Reference*
Orbital energies	1
Population	1
Excited states	1

REFERENCE

1. P.E. Stevenson and W.N. Lipscomb, *J. chem. Phys.* **50**, 3306 (1969).

TiH_3F

State	Geometry		Basis set	Energy	Wave function	Reference
	R(Ti–H)	3.077				
	R(Ti–F)	3.420				
	∠	109° 28′	Min. STO	–947.4642	√	1

Properties	*Reference*
Orbital energies	1
Population analysis	1
Excited states	1

REFERENCE

1. P.E. Stevenson and W.N. Lipscomb, *J. chem. Phys.* **50**, 3306 (1969).

$NiF_6^{=}$

State	Geometry	Basis set	Energy	Wave function	Reference
$^3A_2$					
$^3T_2$		GTO			1, 2

Properties	*References*
Crystal field splitting	1, 2
Hyperfine constants	1, 2

REFERENCES

1. C. Hollister, J.W. Moskowitz, and H. Basch, *Chem. Phys. Lett*, **3**, 185 (1969).

2. C. Hollister, J.W. Moskowitz, and H. Basch, *Chem. Phys. Lett.* **3**, 728 (1969), erratum to ref. 1, which was wrong due to mispunched card.

NiF_6^{4-}

State	Geometry	Basis set	Energy	Wave function	Reference
$^3A_{2g}$	Cell edge 7.585	STO unrestricted Hartree–Fock			1
$^3A_{2g}$		Contr. GTO	–2084.4339	✓	2
$^3T_{2g}$		STO unrestricted Hartree–Fock			1
$^3T_{2g}$		Contr. GTO	–2084.4117		2
$^3T_{1g}$		Contr. GTO	–2084.3418		2
$^1T_{1g}$		Contr. GTO	–2084.2970		2

Properties	*References*
Orbital energies	1, 2
Atomic population	2
Hyperfine constants	1
Crystal effects	1, 2

REFERENCES

1. D.E. Ellis, A.J. Freeman, and P. Ros, *Phys. Rev.* **176**, 688 (1968). Also does calculations on $KNiF_3$ crystal cluster.

2. H.M. Gladney and A. Veillard, *Phys. Rev.* **180**, 385 (1969). Also does calculations on crystal cluster.

$T_cH_9^=$

Properties	*Reference*
Excitation energies	1
Magnetic properties	1

REFERENCE

1. H. Basch and A.P. Ginsberg. *J. phys. Chem., Ithaca,* **73**, 854 (1969). Contracted Gaussian basis set.

bis(π-allyl)Ni

Properties	*References*
Population analysis	1, 2
Orbital energies	1, 2

REFERENCES

1. A. Veillard, *J. chem. Soc. D.*, 1022 (1969).
2. A. Veillard, *J. chem. Soc. D.*, 1427 (1969).
Erratum to ref. 1. Reports energy with GTO basis as –1723.396

$SO_4^=$

Properties	*Reference*
Atomic population	1
Orbital energies	1

REFERENCE

1. I.H. Hillier and V.R. Saunders, *J. chem. Soc. D.*, 1181 (1969).
Min. GTO. energy = –688.327.

$CrO_4^=$

Properties	*Reference*
Atomic population	1
Orbital energies	1

REFERENCE

1. I.H. Hillier and V.R. Saunders, *J. chem. Soc. D.*, 1275 (1969).
Min. GTO.

MnO_4^-

Properties	*Reference*
Atomic population	1
Orbital energies	1

REFERENCE

1. I.H. Hillier and V.R. Saunders, *J. chem. Soc. D.* 1275 (1969).
Min. GTO.

CH_4

State	Geometry	Basis set	Energy	Wave function	Reference
	R(C–H)				
$^1A_1$	2.107	Floating spherical Gaussian	–33.992		37, 44
	2.0	One-centre STO	–38.75		7
	2.0665	GTO + CI	–39.174	√	12
	1.975	Single centre	–39.33		4
	1.971	One-centre STO	–39.349		18
	2.0	One centre	–39.47 (corr)	√	1
	2.0	One-centre STO	–39.62		9
	2.0	One-centre STO	–39.64	√	8
		One centre	–39.68		6
	2.0	One-centre STO + CI	–39.80	√	13
	2.0665	One-centre STO	–39.8444	√	22
	1.994	STO + CI	–39.854	√	28
	2.0	STO	–39.863	√	20
	2.08	One-centre STO	–39.8659	√	16, 23
	2.013	One centre	–39.894	√	21
	2.0	One-centre numerical + CI	–39.90	√	15
	1.99	One-centre STO	–39.94	√	30
	2.053	GTO	–40.06		38
	2.0665	One-centre, ext STO	–40.06563	√	36
	2.0665	Min. SC group function with GTO	–40.098		43
	2.0665	Min. STO	–40.1141	√	32
	2.0665	STO	–40.118		29
	2.05	Min. STO	–40.12822	√	34
		GTO	–40.1428		33
	2.0	STO	–40.15		11
	2.0665	GTO	–40.1668	√	19, 26
	2.040	GTO	–40.172		47
	2.0665	GTO + CI	–40.181	√	24

State	Geometry	Basis set	Energy	Wave function	Reference
	R(C–H)				
$^1A_1$	2.0665	Gaussian lobe natural orbital	–40.1828	√	45
	2.0665	Gaussian lobe	–40.1889	√	46
	2.12	GTO	–40.198		35
	2.0665	STO	–40.20452	√	41

Properties	*References*
Magnetic properties	1, 2, 3, 5, 6, 14, 18, 21, 41
Charge densities	1, 20, 21, 32
Polarizability	1, 21, 40, 41
Spectroscopic constants	6, 10, 13, 18, 21, 22, 34, 38, 47
Ionization potential	7, 11, 12, 18, 23, 38, 47
Orbital energies	9, 12, 19, 20, 23, 24, 26, 27, 34, 36, 41, 46
Dissociation energy	12, 18, 47
Potential curve	13, 21, 24
Quadrupole moment	18
Octopole moment	21, 25, 27, 34, 36
Term values	27
Population analysis	32
Quadrupole coupling constant	34

REFERENCES

1. R.A. Buckingham, H.S.W. Massey, and S.R. Tibbs, *Proc. R. Soc.* **A178**, 119 (1941).

2. C.A. Coulson, *Proc. phys. Soc.* **54**, 51 (1942).
STO basis.

3. H. Hartmann, *Z. Naturf.* **2a**, 489 (1947).
One-centre STO.

4. M.J.M. Bernal, *Proc. phys. Soc.* **A66**, 514 (1953).

5. K.A. Ventratachalam and M.B. Kabadi, *J. phys. Chem., Ithaca*, **59**, 740 (1955).
V.B.

6. C.A. Carter, *Proc. R. Soc.* **A235**, 321 (1956).

7. K. Funabashi and S.L. Magee, *J. chem. Phys.* **26**, 407 (1957).
$^1T_2$ also calculated.

8. S. Koide, H. Sekiyama, and T. Magashima, *J. phys. Soc. Japan.* **12**, 1016 (1957).

9. I. Mills, *Molec. Phys.* **1**, 99 (1958); **4**, 57 (1961). Second paper corrects errors in first.

10. H. Hartmann and G. Gliemann, *Z. phys. Chem.* **15**, 108 (1958). One-centre STO. CD_4 included.

11. S. Besnainou and M. Roux, *J. Chim. phys.* **56**, 250 (1959).

12. R.K. Nesbet, *J. chem. Phys.* **32**, 1114 (1960). Correlation by perturbation method.

13. A.F. Saturno and R.G. Parr, *J. chem. Phys.* **33**, 22 (1960).

14. K.E. Banyard, *J. chem. Phys.* **33**, 832 (1960). Wave functions as in ref. 4.

15. E.L. Albasiny and I.R.A. Cooper, *Molec. Phys.* **4**, 353 (1961).

16. R. Moccia, *J. chem. Phys.* **37**, 910 (1962).

17. J.R. Hoyland and F.W. Lampe, *J. chem. Phys.* **37**, 1066 (1962). Proton affinity using one-centre STO.

18. F. Grein. *Theor. Chim. Acta* **1**, 52 (1962).

19. M. Krauss, *J. chem. Phys.* **38**, 564 (1963).

20. J.J. Sinai, *J. chem. Phys.* **39**, 1575 (1963).

21. E.L. Albasiny and I.R.A. Cooper, *Proc. phys. Soc.* **82**, 289 (1963).

22. D.M. Bishop, *Molec. Phys.* **6**, 305 (1963). Non-integral quantum numbers.

23. R. Moccia, *J. chem. Phys.* **40**, 2164 (1964).

24. B.J. Woznik, *J. chem. Phys.* **40**, 2860 (1964).

25. J.J. Sinai, *J. chem. Phys.* **40**, 3596 (1964). Wave functions as in refs 19, 20, 24.

26. M. Krauss, *J. Res. natn. Bur. Stand.* **68A**, 635 (1964).

27. M. Klessinger and R. McWeeny, *J. chem. Phys.* **42**, 3343 (1965). SC group functions with min. STO. Wave functions given.

28. R.M. Rutledge and A.F. Saturno, *J. chem. Phys.* **44**, 977 (1966).

29. B. Kochel, *Z. Naturf.* **18a**, 739 (1963); **20a**, 1472 (1965).

30. R.M. Rutledge and A.F. Saturno, *J. chem. Phys.* **44**, 977 (1966).

31. T. Caves and M. Karplus, *J. chem. Phys.* **45**, 1670 (1966). Deuteron electron field gradient and quadrupole coupling constant in CH_3D given. Wave functions as in ref. 24.

32. W.E. Palke and W.N. Lipscomb, *J. Am. chem. Soc.* **88**, 2384 (1966).

33. L.C. Snyder, *J. chem. Phys.* **46**, 3602 (1967).

34. R.M. Pitzer, *J. chem. Phys.* **46**, 4871 (1967).

35. C.D. Ritchie and H.F. King, *J. chem. Phys.* **47**, 564 (1967).
Correlation energy calculated.

36. J.R. Hoyland, *J. chem. Phys.* **47**, 3556 (1967).

37. A.A. Frost, B.H. Prentice, and R.A. Rouse, *J. Am. chem. Soc.* **89**, 3064 (1967).

38. R. Janoschek, G. Diercksen, and H. Preuss, *Int. J. Quantum Chem.* **1**, 373 (1967).

39. P. Pyykko, *Proc. phys. Soc.* **92**, 841 (1967).
Deutron quadrupole coupling constant. Wave functions as in refs. 15 and 23.

40. G.P. Arrighini, M. Maestro and R. Moccia, *Chem. Phys. Lett.* **1**, 242 (1967).
Min. STO and double zeta basis.

41. G.P. Arrighini, C. Guidotti, M. Maestro, R. Moccia, and O. Salvetti, *J. chem. Phys.* **49**, 2224 (1968).

42. G.P. Arrighini, M. Maestro, and R. Moccia, *Symp. Faraday Soc.* **2** 48 (1968).
Dipole hyperpolarizability using STO basis.

43. M. Klessinger, *Symp. Faraday Soc.* **2**, 73 (1968).

44. A.A. Frost, *J. phys. Chem., Ithaca,* **72**, 1289 (1968).

45. R. Ahlrichs and W. Kutzelrigg, *Chem. Phys. Lett.* **1**, 651 (1967).

46. S. Rothenberg, *J. chem. Phys.* **51**, 3389 (1969).

47. H. Preuss and R. Janoschek, *J. molec Struct.* **3**, 423 (1969).

CH_5^+

State	Geometry	Basis set	Energy	Wave function	Reference
$^1A_1'$	D_{3h}	STO + CI	−40.0363	√	5
		GTO	−40.2842		7
		GTO	−40.3351		6
			−40.3523		8
$^1A_1$	C_{3v}	STO + CI	−40.0382	√	5
		GTO	−40.1854		6
$^1A_1$	C_{4v}	GTO	−40.2892		7
		GTO	−40.3333		6
			−40.3452		8

Properties	*References*
Potential surface	2, 4
Dissociation energy	4
Orbital energies	5

REFERENCES

1. J.R. Hoyland and F.W. Lampe, *J. chem. Phys.* **37**, 1066 (1957). One-centre STO. Relative energies of CH_3^+, CH_3, CH_4, CH_5^+ given.

2. J. Higuchi, *J. chem. Phys.* **31**, 563 (1959). D_{3h} symmetry. Proton affinity of CH_4 calculated using wave function in M.J.M. Bernal, *Proc. phys. Soc.* **A66**, 514 (1953).

3. H. Hartmann and F. Grein, *Z. phys. Chem.* **22**, 305 (1959). One-centre calculation of energy for CH_4 proton affinity.

4. M. Yamazaki, *J. phys. Soc. Japan* **14**, 498 (1959). $^1A_1'$ state in D_{3h} configuration. VB wave function given. Proton affinity of CH_4 calculated.

5. M. Rutledge and F. Saturno, *J. chem. Phys.* **43**, 597 (1965). Coordinates given. C_{4v} included but no energy given.

6. J.L. Gole, *Chem. Phys. Lett.* **3**, 577 (1969). Bond lengths etc. given in A. Gamba, G. Merosi and M. Sunonetta, *Chem. Phys. Lett.* **3**, 20 (1969).

7. W. Th. A.M. van der Lugt and P. Ros, *Chem. Phys. Lett.* **4**, 389 (1969). Geometry given in paper.

8. J.L. Gole, *Chem. Phys. Lett.* **4**, 408 (1969). No details of geometry or basis set.

CH_5^-

State	Geometry	Basis set	Energy	Wave function	Reference
$^1A_1'$	D_{3h}	GTO	–40.4659		1
$^1A_1$	C_{4v}	GTO	–40.3793		1

REFERENCE

1. W. Th. A.M. van der Lugt and P. Ros, *Chem. Phys. Lett.* **4**, 389 (1969). Geometry given.

C_2H_6

State	Geometry	Basis set	Energy	Wave function	Reference
$^1A_{1g}$	Experimental	GTO	−78.57040		9
	Coordinates given	Min. STO	−78.99115	✓	1
	Coordinates given	Min. STO	−79.0689	✓	3
	Experimental	Min. STO	−79.09797	✓	11
	R(C–C) 2.915878 R(C–H) 2.0825	Contracted GTO	−79.1025	✓	5
	Coordinates given	GTO lobe	−79.14778		8
	R(C–C) 2.9102 R(C–H) 2.0975	GTO lobe	−79.1825	✓	7
	R(C–C) 2.916	GTO	−79.2377		15

Properties	*References*
Charge density	1, 3, 9, 11, 13
Orbital energies	5, 11
Barrier to rotation	1, 2, 8, 9, 11, 15, 16
Quadrupole moment	4
Spin–spin coupling constant	10, 14
Localized orbitals	17

REFERENCES

1. R.M. Pitzer and W.N. Lipscomb, *J. chem. Phys.* **39**, 1995 (1963).
2. R.E. Wyatt and R.G. Parr, *J. chem. Phys.* **43**, S217 (1965). *J. chem. Phys.* **44**, 1529 (1966).
3. W.E. Palke and W.N. Lipscomb, *J. Am. chem. Soc.* **88**, 2384 (1966).
4. O.J. Sovers, M. Karplus, and C.W. Kern, *J. chem. Phys.* **45**, 3895 (1966).
5. E. Clementi and D.R. Davis, *J. chem. Phys.* **45**, 2593 (1966).
6. U. Kaldor, *J. chem. Phys.* **46**, 1981 (1967). Localized orbitals.
7. R.J. Buenker, S.D. Peyerimhoff, and J.L. Whitten, *J. chem. Phys.* **46**, 2029 (1967).
8. W. Fink and L.C. Allen, *J. chem. Phys.* **46**, 2261 (1967). Potential curves.
9. L. Pedersen and K. Morokuma, *J. chem. Phys.* **46**, 3941 (1967).
10. C. Barbier and G. Berthier, *Int. J. Quantum Chem.* **1**, 657 (1967).
11. R.M. Pitzer, *J. chem. Phys.* **47**, 965 (1967).

12. M. Klessinger, *Symp. Faraday Soc.* **2**, 73 (1968).
Group functions.

13. A.A. Frost and R.A. Rouse, *J. Am. chem. Soc.* **90**, 1965 (1968).
Floating spherical gaussian calculation.

14. T. Yonezawa, I. Morishima, M. Fujii, and H. Kato, *Bull. chem. Soc. Japan* **42**, 1248 (1969).

15. A. Veillard, *Chem. Phys. Lett.* **3**, 128 (1969).

16. A. Veillard, *Chem. Phys. Lett.* **3**, 565 (1969).
Rotational barrier.

17. S. Rothenberg, *J. chem. Phys.* **51**, 3389 (1969).
Localized orbitals.

$C_2H_6^+$

State	Geometry	Basis set	Energy	Wave function	Reference
$^2A_1$	R(C–C) 3.066	GTO	–78.9360		1

Property	*Reference*
Rotational barrier	1

REFERENCE

1. A. Viellard, *Chem. Phys. Lett.* **3**, 128 (1969).

$C_2H_6^{++}$

1. S.D. Peyerimhoff and R.J. Buenker, *J. chem. Phys.* **49**, 312 (1968).
Stabilities of bridged structures discussed.

C_3H_8

State	Geometry	Basis set	Energy	Wave function	Reference
^1A	Staggered	Gaussian lobe	–117.438		1
	Eclipsed	Gaussian lobe	–117.425		1
	R(C–C) 2.92 R(C–H) 2.08	GTO	–118.065		2

Property	*References*
Rotation barrier	1, 2

REFERENCES

1. J.R. Hoyland, *Chem. Phys. Lett.* **1**, 247 (1967).
2. J.R. Hoyland, *J. chem. Phys.* **49**, 1908 (1968). Coordinates given.

C_4H_{10}

REFERENCE

1. J.R. Hoyland, *J. chem. Phys.* **49**, 2563 (1968). GTO Study of internal rotation. $E = -157.031573$. Charge density.

C_2H_4

State	Geometry			Basis set	Energy	Wave function	Reference
$^1A_{1g}$	R(C–C)	R(C–H)	∠				
	2.255	2.02		Contracted GTO	−76.77		13
	2.562	2.11		GTO	−77.5050		28, 29
				GTO	−77.8002		2
				Min. STO	−77.8343	✓	6
	2.5	2.0	120°	STO	−77.84116		16
				STO	−77.8558		24
				STO + CI	−77.8758	✓	19
	2.5510	2.0236	117°	GTO lobe	−77.93725		24
				GTO	−77.95022		3
	2.55102	2.0236	117°	GTO	−78.0012		4, 11
	Experimental			GTO	−78.0062		7
	Experimental			Ext. GTO	−78.0140		22
$^3B_{3u}$	Ground state geom.			Ext. GTO	−77.8917		22
$^1B_{3u}$	Ground state geom.			Ext. GTO	−77.7415		22

Properties	*References*
Orbital energies	2, 4, 5, 7, 12, 16, 19, 24
Charge density	6, 17, 22
Excitation energies	2, 5, 7, 12, 19, 23, 25
Spin coupling constants	8, 10, 21

Properties	*References*
Spectroscopic constants	2, 5, 12, 22
Ionization potential	13, 19
Potential surface	20, 28
Barrier to rotation	20
Dipole moment	24
Oscillator strength	27
Population analysis	28

REFERENCES

1. C.R. Hueller, *J. chem. Phys.* **22**, 120 (1954).
   Approximate calculation.
2. J.W. Moskowitz and M.C. Harrison, *J. chem. Phys.* **42**, 1726 (1965).
3. J.W. Moskowitz, *J. chem. Phys.* **43**, 60 (1965).
4. J.L. Whitten, *J. chem. Phys.* **44**, 359 (1966).
5. M.B. Robin, R.R. Hart and N.A. Kuebler, *J. chem. Phys.* **44**, 1803 (1966).
   Discuss excited states.
6. W.E. Palke and W.N. Lipscomb, *J. Am. chem. Soc.* **88**, 2384 (1966).
7. J.M. Schulman, J.W. Moskowitz, and C. Hollister, *J. chem. Phys.* **46**, 2759 (1967).
8. E.A.G. Armour and A.J. Stone, *Proc. R. Soc.* **A302**, 25 (1967).
9. U. Kaldor, *J. chem. Phys.* **46**, 1981 (1967).
   Localized orbitals.
10. C. Barbier and G. Berthier, *Int. J. Quantum Chem.* **1**, 657 (1967).
11. R.J. Buenker, S.D. Peyerimhoff, and J.L. Whitten, *J. chem. Phys.* **46**, 2029 (1967).
12. T.H. Dunning and V. McKay, *J. chem. Phys.* **47**, 1735 (1967).
    Includes excited states.
13. G. Diercksen and H. Preuss, *Int. J. Quantum Chem.* **1**, 365 (1967).
14. H. Preuss and G. Diercksen, *Int. J. Quantum Chem.* **1**, 369 (1967).
    1s orbital energy.
15. A.A. Frost, B.H. Prentice, and R.A. Rouse, *J. Am. chem. Soc.* **89**, 3064 (1967).
    Floating spherical gaussian calculation.
16. P. Rajagopal, *Z. Naturf.* **22a**, 295 (1967).
17. A.A. Frost and R.A. Rouse, *J. Am. chem. Soc.* **90**, 1965 (1968).
    Floating spherical gaussian calculation.
18. M. Klessinger, *Sym. Faraday Soc.* **2**, 73 (1968).
    Group orbitals.

19. U. Kaldor and I. Shavitt, *J. chem. Phys.* **48**, 191 (1968).
Includes excited states and ions.

20. R.J. Buenker, *J. chem. Phys.* **48**, 1368 (1968).
Barrier to rotation.

21. T. Yonezawa, I. Morishima, M. Fujii, and H. Kato, *Bull. chem. Soc. Japan* **42**, 1248 (1969).

22. T.H. Dunning, W.J. Hunt, and W.A. Goddard, *Chem. Phys. Lett.* **4**, 147 (1969).

23. S. Huzinaga, *Theor. Chim. Acta* **15**, 12 (1969).
Excited states.

24. J.D. Petke and J.L. Whitten, *J. chem. Phys.* **51**, 3166 (1969).

25. A.F. Hausen, *Theor. Chim. Acta* **14**, 363 (1969).
Excitation energies by valence bond calculation.

26. E. Switkes, R.H. Stevens, and W.N. Lipscomb, *J. chem. Phys.* **51**, 5229 (1969).
Min. STO.

27. K.J. Miller, *J. chem. Phys.* **51**, 5235 (1969).
Oscillator strengths.

28. P. Swanstrom, R. Janoschek, and H. Preuss. *Int. J. Quantum Chem.* **3**, 873 (1969).

29. H. Preuss and R. Janoschek, *J. molec. Struct.* **3**, 423 (1969).

$CH_2{=}CH^{\cdot}$

State	Geometry		Basis set	Energy	Wave function	Reference
Ground state	R(C–C)	2.53				
	R(C–H)	2.04				
	$\hat{HCH}$	120°				
	$\hat{CCH}$	137°	GTO	–77.3620		1

Property	*Reference*
Population analysis	1

REFERENCE

1. P. Millie and G. Berthier, *Int. J. Quantum Chem.* **2S**, 67 (1968).

$CH_3-CH=CH_2$

State	Geometry	Basis set	Energy	Wave function	Reference
Ground state	Staggered	GTO	–116.39646		1
^1A	Eclipsed		–116.39408		1

Properties	*Reference*
Orbital energies	1
Charge densities	1
Population analysis	1
Dipole moment	1
Quadrupole moment	1
Magnetic properties	1

REFERENCE

1. M.L. Unland, J.R. van Wazer, and J.H. Letcher, *J. Am. chem. Soc.* **91**, 1045 (1969).
Rotational barrier calculated. Coordinates given.

Allene C_3H_4

State	Geometry	Basis set	Energy	Wave function	Reference
$^1A_1$	R(C–H) 2.054 R(C–C) 2.462 $\hat{HCH}$ 118°	GTO	–114.8392		2
	Coords given	GTO	–115.039		4
	Coords given	GTO	–115.3110		3
	As above	Gaussian lobe + CI	–115.6979		1

Properties	*References*
Orbital energies	1, 2, 3
Potential surface	1
Ionization potential	2
Excitation energies	2
Population analysis	3

REFERENCES

1. R.J. Buenker, *J. chem. Phys.* **48**, 1368 (1968).

2. L.J. Schaad, L.A. Burnelle, and K.P. Dressler, *Theor. Chim. Acta* **15**, 91 (1969).
Virtual orbitals for excited states.

3. J.M. André, M. Cl. André, G. Leroy, and J. Weiler, *Int. J. Quantum Chem.* **3**, 1013 (1969).

4. H. Preuss and R. Janoschek, *J. molec Struct.* **3**, 423 (1969).

Allyl cation

State	Geometry R(C–C)	Geometry R(C–H)	Basis set	Energy	Wave function	Reference
Ground state	2.686	2.04	GTO	–111.73443		1

Properties	*Reference*
Orbital energies	1
Charge densities	1
Population analysis	1
Potential surface	1

REFERENCE

1. D.T. Clark and D.R. Armstrong, *Theor. Chim. Acta* **13**, 365 (1965).
Reaction surface for cyclopropyl cation ⇌ allyl cation.

Allyl anion

State	Geometry	Basis set	Energy	Wave function	Reference
Ground state	See allyl cation (ref. 1)	GTO	–111.94671		1

Properties	*Reference*
Charge density	1
Population analysis	1

REFERENCE

1. D.T. Clark and D.R. Armstrong, *Theor. Chim. Acta* **14**, 376 (1969).
Reaction surface for cyclopropyl anion ⇌ allyl anion.

C_4H_8

State	Geometry	Basis set	Energy	Wave function	Reference
trans	Coords given	GTO	−153.3490		1
cis			−153.3472		1

Properties	*References*
Ionization potentials	1, 2
Potential curve	1
Rotational barrier	1, 2

REFERENCES

1. R. Janoschek and H. Preuss, *Int. J. Quantum Chem.* **3**, 873 (1969).
2. H. Preuss and R. Janoschek, *J. Molec. Struct.* **3**, 423 (1969).

C_4H_6 butadiene

State	Geometry	Basis set	Energy	Wave function	Reference
^1A_g	*Trans* (exp.)	Gaussian lobe + CI	−154.7103		1
$^1A_1$	*Cis* (exp.)	Gaussian lobe + CI	−154.7023		1

Properties	*Reference*
Orbital energies	1
Excitation energies	1
Ionization potentials	1

REFERENCE

1. R.J. Buenker and J.L. Whitten, *J. chem. Phys.* **49**, 5381 (1968). Excited states ^3B_u, ^3A_g, ^1B_u and ^1A_s discussed.

$CH_3 - C \equiv C - H$, C_3H_4

State	Geometry	Basis set	Energy	Wave function	Reference
Ground State		Min. STO	−115.5830	/	1

Properties	*Reference*
Ionization potentials	1
Dipole moment	1
Orbital energies	1
Population analysis	1

REFERENCES

1. M.D. Newton and W.N. Lipscomb, *J. Am. chem. Soc.* **89**, 4261 (1967).
Coordinates given from
$R(H_1, C_1) = 1.996$ Tetrahedral angles
$R(C_1, C_2) = 2.279$
$R(C_2, C_3) = 2.758$
$R(C_3, H_{2,3,4}) = 2.083$

Carbonium Ions

State	Basis set	Energy	Wave function	Reference
$C_2H_5^+$				
Ground state	GTO	–77.779		2, 3
	GTO	–78.2422		1
		Protonated ethylene		
$CH_2CH_3^+$				
Ground state	GTO	–78.2279		1
		Vinyl$^+$		
CH_2-CH^+				
Ground state	GTO	–76.9847		1
		Protonated acetylene		
$HC{\equiv}CH_2^+$				
Ground state	GTO	–76.9447		1

REFERENCES

1. R. Sustman, J.E. Williams, M.J.S. Dewar, L.C. Allen, and P. von R. Schlayer, *J. Am. chem. Soc.* **91**, 5350 (1969). Geometries given in paper.

2. F. Frater, R. Janoschek and H. Preuss, *Int. J. Quantum Chem.* **3**, 873 (1969).

3. H. Preuss and R. Janoschek, *J. molec. Struct.* **3**, 423 (1969).

CH_3NH_2

State	Geometry	Basis set	Energy	Wave function	Reference
Ground state $^1A'$	Staggered (Exp.)	GTO	−94.32348		
		Gaussian lobe	−95.11273	✓	1
	Eclipsed (Exp.)	Gaussian lobe	−95.10890	✓	1

Properties	*References*
Population analysis	2
Nuclear quadrupole coupling	3

REFERENCES

1. W.H. Fink and L.C. Allen, *J. chem. Phys.* **46**, 2276 (1967). Rotational barrier calculated. Coordinates given.
2. L. Pedersen and K. Morokuma, *J. chem. Phys.* **46**, 395 (1967). Rotational barrier calculated.
3. C.T. O'Konski and T-K. Ha, *J. chem. Phys.* **49**, 5354 (1968). Wave functions as in ref. 1.

CH_3OH

State	Geometry	Basis set	Energy	Wave function	Reference
Ground state ^1A	Staggered (Exp.)	GTO	−113.90470		2
		Min. Gaussian lobe	−114.9343	✓	3
		Gaussian lobe	−114.99574	✓	1
Ground state	Eclipsed (Exp.)	Min. Gaussian lobe	−114.93212	✓	3
		Gaussian lobe	−114.99405	✓	1

Properties	*References*
Orbital energies	1, 3
Potential curve	1
Potential analysis	2
Rotation barrier	1, 2, 3

REFERENCES

1. W.H. Fink and L.C. Allen, *J. chem. Phys.* **46**, 2261 (1967). Coordinates given. Potential barrier calculated.
2. L. Pedersen and K. Morokuma, *J. chem. Phys.* **46**, 3941 (1967). Potential barrier calculated.
3. S. Rothenberg, *J. chem. Phys.* **51**, 3389 (1969).

CH_3F

State	Geometry		Basis set	Energy	Wave function	Reference
$^1A_1$	R(C–H)	2.04	Min. GTO	−138.2500		3
	R(C–F)	2.68	GTO	−139.0342		
	∠	107°30′				
	R(C–H)	2.067				
	∠	109°28′	Ext. STO	−139.0612	√	4

Properties	*Reference*
Dipole hyperpolarizability	2
Orbital energies	3, 4
Electrical properties	4
Magnetic properties	4

REFERENCES

1. M. Klessinger, *Symp. Faraday Soc.* **2**, 73 (1968). Self-consistent group functions.
2. G.P. Arrighini, M. Maestro, and R. Moccia, *Symp. Faraday Soc.* **2** (1968). Dipole hyperpolarizability by perturbed Hartree–Fock.
3. G. Berthier, D-J David, and A. Veillard, *Theor. Chim. Acta* **14**, 329 (1969). Calculates reaction surface, activation energy for $F^- + CH_3F \rightarrow CH_3F + F^-$.

4. G.P. Arrighini, G. Guidotti, M. Maestro, R. Moccia, and O. Salvati, *J. chem. Phys.* **51**, 480 (1969).
Many one-electron props calculated.

CH_2N_2 diazomethane

State	Geometry	Basis set	Energy	Wave function	Reference
$^1A_1$	Coordinates given	Min. GTO	–147.0100		1

Properties	*Reference*
Charge density	1
Orbital energies	1
Population analysis	1
Dipole moment	1

REFERENCE

1. J.M. André, M. Cl. André, G. Leroy, and J. Weiler, *Int. J. Quantum Chem.* **3**, 1013 (1969).

HCOOH

State	Geometry	Basis set	Energy	Wave function	Reference
$^1A'$	Experimental	GTO	–183.2572		1
		Gaussian lobe	–188.6552		2

Properties	*References*
Potential surface	1, 2
Ionization potential	1
Population analysis	1
Orbital energies	2
Charge densities	2
Dipole moment	2

REFERENCES

1. P. Ros, *J. chem. Phys.* **49**, 4902 (1968).
Protonated form included. Coordinates given.

2. S.D. Peyerimhoff and R.J. Buenker, *J. chem. Phys.* **50**, 1846 (1969).
Excited states discussed. CI included. $HCOO^-$ also calculated.

$HCOO^-$

State	Geometry	Basis set	Energy	Wave function	Reference
$^1A_1$	Experimental	Gaussian lobe + CI	–188.1132		1

Properties	*Reference*
Orbital energies	1
Charge densities	1
Potential surface	1

REFERENCE

1. S.D. Peyerimhoff, *J. chem. Phys.* **47**, 349 (1967).
Excited states discussed.

$HCONH_2$

State	Geometry	Basis set	Energy	Wave function	Reference
^1A	Experimental	GTO	–168.5259		2
	Experimental	GTO	–168.8684		1
	Experimental	GTO	–168.872		3

Properties	*References*
Orbital energies	1, 2, 3
Excitation energies	1
Population analysis	1, 2
Dipole moment	1, 2, 3
Ionization potential	2
Charge densities	2
Barrier to rotation	1, 2, 3

REFERENCES

1. H. Basch, M.B. Robin, and N.A. Kuebler, *J. chem. Phys.* **47**, 1201 (1967).
Coordinates given. Excited states included.

2. M.A. Robb and I.G. Csizmadia, *Theor. Chim. Acta* **10**, 269 (1968).
Low-lying excited states.

3. M.A. Robb and I.G. Csizmadia, *J. chem. Phys.* **50**, 1819 (1969).
Tautomers also treated.

H_2COH^+

State	Geometry	Basis set	Energy	Wave function	Reference
Ground state	See below	Min. GTO	–110.88722		2
	See below	GTO	–113.8609		1

Properties	*Reference*
Potential curve	2
Charge densities	2

REFERENCES

1. A.C. Hopkinson, N.K. Holbrook, K. Yates, and I.G. Csizmadia, *J. chem. Phys.* **49**, 3596 (1968).
   Geometry $R(\text{C–O}) = 2.419$, $\hat{\text{HCH}} = 118°$
   $R(\text{C–H}) = 2.117$, $\hat{\text{COH}} = 120°$
   $R(\text{O–H}) = 1.814$, Planar
   Proton affinity of H_2CO calculated with different basis sets.
2. P. Ros, *J. chem. Phys.* **49**, 4902 (1968).
   Geometry $R(\text{C–O}) = 2.4$, $\hat{\text{HCH}} = 120°$
   $R(\text{C–H}) = 2.06$, $\hat{\text{COH}} = 120°$
   $R(\text{O–H}) = 1.86$, Planar
   Potential curve as function of $\hat{\text{COH}}$ given. Stabilities of various conformers included.

CH_3CHO

State	Geometry	Basis set	Energy	Wave function	Reference
Ground state $^1A'$	Experimental	GTO	–148.4999		1

Properties	*Reference*
Ionization potential	1
Overlap population	1
Rotation barrier	1

REFERENCE

1. P. Ros, *J. chem. Phys.* **49**, 4902 (1968).
   Protonated form included. Rotational barrier calculated. Coordinates given.

Ketene C_2H_2O

State	Geometry	Basis set	Energy	Wave function	Reference
Ground state $^1A_1$	Experimental	Min GTO	−150.9177		2
		GTO	−151.507701	√	1

Properties	*References*
Orbital energies	1, 2
Charge densities	1, 2
Population analysis	1, 2
Dipole moment	1, 2
Quadrupole moment	1
Magnetic properties	1

REFERENCES

1. J.H. Letcher, M.L. Unland, and J.R. van Wazer, *J. chem. Phys.* **50**, 2185 (1969).

2. J.M. André, M. Cl. André, G. Leroy, and J. Weiller, *Int. J. Quantum Chem.* **3**, 1013 (1969) .

NCCCH

State	Geometry	Basis set	Energy	Wave function	Reference
$^1\Sigma^+$	R(N–C) 2.1864				
	R(C–C) 2.6116				
	R(C–C) 2.2734				
	R(C–H) 1.9975	Ext. STO	−168.5784	√	1, 2

Properties	*References*
Orbital energies	1, 3
Dipole moment	2, 3, 4
Dissociation energy	2
Quadrupole coupling constant	4
Charge densities	4

REFERENCES

1. A.D. McLean and M. Yoshimine, Supp. to *I.B.M. Jl. Res. Dev.* (1967). Hartree–Fock limit.

2. M. Yoshimine and A.D. McLean, *Int. J. Quantum Chem.* **1S**, 313 (1967).
Hartree–Fock limit.

3. E. Clementi and D. Klint, *J. chem. Phys.* **50**, 4899 (1969).
Wave-functions as in ref. 1.

4. R. Bonaccorsi, E. Scrocco, and J. Tomasi, *J. chem. Phys.* **50**, 2940 (1969).
Wave functions as in ref. 1.

Vinyl Cyanide CH_2CHCN

State	Geometry	Basis set	Energy	Wave function	Reference
$^1A'$	Experimental	GTO	–160.690761		1

Properties	*Reference*
Orbital energies	1

REFERENCE

1. J.B. Moffat and R.J. Collens, *J. molec. Spectrosc.* **27**, 252 (1968).
Coordinates given.

Nitriles

Property	*Reference*
Magnetic susceptibility	1

REFERENCE

1. J. Baudet, J. Tillieu and J. Guy, *C.r. hebd. Séanc. Acad. Sci. Paris* **244**, 1756 (1956).
Acetonitrile, propionitrile, butyronitrile, and iso amylcyanide treated using STO basis.

CH_3SOH

Properties	*Reference*
Rotational barrier	1
Orbital energies	1
Atomic population	1
Charge density	1

REFERENCE

1. A. Rauk, S. Wolfe, and I.G. Csizmadia, *Can. J. Chem.* **47**, 113 (1969). GTO energy = –488.1516.

CH_2SOH^-

Properties	*Reference*
Rotational barrier	1
Orbital energies	1
Atomic population	1
Charge density	1

REFERENCE

1. A. Rauk, S. Wolfe and I.G. Csizmadia, *Can. J. Chem.* **47**, 113 (1969). GTO energy = 487.2979.

$(CH_3)_2$ SO dimethyl sulphoxide

1. A. Rauk, S. Wolfe, and I.G. Csizmadia, *Can. J. Chem.* **47**, 113 (1969). Rotational barrier.

$CH_3CH_2SO^-$

REFERENCE

1. A. Rauk, S. Wolfe, and I.G. Csizmadia, *Can. J. Chem.* **47**, 113 (1969).

Cyclopropane

State	Geometry		Basis set	Energy	Wave function	Reference
	R(C–C)	*R*(C–H)				
$^1A_1$	2.847	2.108	Floating Spherical Gaussian	–98.895		3
	2.419		GTO	–115.9978		7
	2.91		GTO	–116.02		1, 9
			GTO	–116.6797		5
			Gaussian lobe	–116.91636		4
			GTO	–116.9918		6
			GTO	–117.0099		8

Properties	*References*
Ionization potential	1, 6, 9
Reaction surface	1, 9
Charge density	3, 6
Orbital energies	6, 7, 8
Population analysis	6, 7
Magnetic properties	8

REFERENCES

1. H. Preuss and G. Diercksen, *Int. J. Quantum Chem.* **1**, 361 (1967).
2. H. Preuss and G. Diercksen, *Int. J. Quantum Chem.* **1**, 369 (1967). 1s orbital energy discussed.
3. A.A. Frost and R.A. Rouse, *J. Am. chem. Soc.* **90**, 1965 (1968). Coordinates given.
4. J.D. Petke and J.L. Whitten, *J. Am. chem. Soc.* **90**, 3338 (1968).
5. M. Klessinger, *Symp. Faraday Soc.* **2**, 73 (1968).
6. L. Kochanski and J.M. Lehn, *Theor. Chim. Acta* **14**, 281 (1969).
7. D.T. Clark, *Theor. Chim. Acta* **15**, 225 (1969).
8. H. Basch, M.B. Robin, N.A. Kuebler, C. Baker, and D.W. Turner, *J. chem. Phys.* **51**, 52 (1969).
9. H. Preuss and R. Janoschek, *J. molec. Struct.* **3**, 423 (1969).

Protonated Cyclopropane $C_3H_7^+$

State	Geometry	Basis set	Energy	Wave function	Reference
	$C\hat{C}C$				
	60°	Gaussian lobe	−117.11375		1
	80°	Gaussian lobe	−117.16159		1

Properties	*Reference*
Charge density	1
Orbital charges	1
Potential surface	1

REFERENCE

1. J.D. Petke and J.L. Whitten, *J. Am. chem. Soc.* **90**, 3338 (1968). Coordinates given.

Cyclopropyl cation $C_3H_5^+$

State	Geometry	Basis set	Energy	Wave function	Reference
	R(C–C) R(C–H)				
	2.859 2.079				
	H$\hat{C}$H				
	116°				
	Planar				
	C_1–H	GTO	–111.63264		1
	Bent				
	C_1–H	GTO	–111.56947		1

Properties	*References*
Orbital energies	1
Charge densities	1
Population analysis	1
Potential surface	1, 2

REFERENCES

1. D.T. Clark and D.R. Armstrong, *Theor. Chim. Acta* **13**, 365 (1969). Reaction surface for cyclopropyl cation ⇌ allyl cation given.

2. D.T. Clark and D.R. Armstrong, *J. chem. Soc. D.*, 850 (1969). Discussion of aromaticity and pseudo aromaticity.

Cyclopropyl anion

State	Geometry	Basis set	Energy	Wave function	Reference
	Bent (exp.)				
	C_1–H	GTO	–111.84286		1
	Planar				
	C_1–H		–111.84022		1

Properties	*References*
Charge density	1
Population analysis	1
Potential surface	2

REFERENCES

1. D.T. Clark and D.R. Armstrong, *Theor. Chim. Acta* **14**, 370 (1969). Geometry in ref. 1 of cyclopropyl cation. Reaction surface for cyclopropyl anion ⇌ allyl anion given.

2. D.T. Clark and D.R. Armstrong, *J. chem. Soc. D.*, 850 (1960). Discussion of aromaticity and pseudo-aromaticity.

Cyclopropene C_3H_4

State	Geometry	Basis set	Energy	Wave function	Reference
$^1A_1$	Experimental	GTO	–114.7725		2
		GTO	–115.7572		1
		GTO	–115.7655		3

Properties	*References*
Orbital energies	1, 2, 3
Charge densities	1
Population analysis	1, 2
Dipole moment	1, 2
Spectroscopic constants	3

REFERENCES

1. L. Kochanski and J.M. Lehn, *Theor. Chim. Acta* **14**, 281 (1969).
2. D.T. Clark, *Theor. Chim. Acta* **15**, 225 (1969).
3. M.B. Robin, H. Basch, N.A. Kuebler, K. Wiberg, and B. Ellison, *J. chem. Phys.* **51**, 45 (1969).

Cyclopropenyl Ions

Property	*Reference*
Potential surface	1

REFERENCE

1. D.T. Clark, *J. chem. Soc. D.*, 637 (1969).
Anti-aromaticity of anion and aromaticity of cation discussed GTO basis.

Azridine C_2H_5N

State	Geometry	Basis set	Energy	Wave function	Reference
$X\ ^1A'$	Experimental	GTO	–131.8049		3
		GTO	–132.9487		1, 2
		GTO	–132.9726		5

Properties	*References*
Dipole moment	1; 2, 3, 5
Ionization potential	2
Potential surface	2
Population analysis	2, 3
Orbital energies	3, 5
Quadrupole coupling constant	4
Magnetic properties	5

REFERENCES

1. A. Veillard, J.M. Lehn, and B. Munsen, *Theor. Chim. Acta.* **9**, 275 (1968).
   N inversion barrier given.
2. J.M. Lehn, B. Munsch, P. Millie, and A. Veillard, *Theor. Chim. Acta* **13**, 313 (1969).
   N inversion barrier given. Coordinates given.
3. D.T. Clark, *Theor. Chim. Acta* **15**, 225 (1969).
4. L. Kochanski, J.M. Lehn, and B. Levy, *Chem. Phys. Lett.* **4**, 75 (1069).
   GTO basis.
5. H. Basch, M. Robin, N.A. Kuebler, C. Baker, and D.W. Turner, *J. chem. Phys.* **51**, 52 (1969).

Oxirane C_2H_4O

State	Geometry	Basis set	Energy	Wave function	Reference
X^1A_1	Experimental	GTO	–151.3951		1
		GTO	–152.6745		3
		GTO	–152.8012		2

Properties	*References*
Dipole moment	1, 2
Population analysis	1
Orbital energies	1, 2
Magnetic properties	2

REFERENCES

1. D.T. Clark, *Theor. Chim. Acta* **15**, 225 (1969).

2. H. Basch, M.B. Robin, N.A. Kuebler, C. Baker and D.W. Turner, *J. chem. Phys.* **51**, 52 (1969).
$^2B_2$, $^2A_1$, and $^2B_1$ states of ion.

3. E.F. Hayes, *J. chem. Phys.* **51**, 4787 (1969).
Geometry given.

Thi-irane C_2H_4S

State	Geometry	Basis set	Energy	Wave function	Reference
$^1A_1$	Experimental	GTO	–456.0016		1

Properties	*Reference*
Dipole moment	1
Orbital energies	1
Population analysis	1

REFERENCE

1. D.T. Clark, *Theor. Chim. Acta* **15**, 225 (1969).

1–Azirene C_2H_3N

State	Geometry	Basis set	Energy	Wave function	Reference
$^1A'$	Experimental	GTO	–130.5935		1

Properties	*Reference*
Dipole moment	1
Orbital energies	1
Population analysis	1

REFERENCE

1. D.T. Clark, *Theor. Chim. Acta* **15**, 225 (1969).

2–Azirene C_2H_3N

State	Geometry	Basis set	Energy	Wave function	Reference
$^1A'$	Experimental	GTO	–130.5478		1

Properties	*Reference*
Dipole moment	1
Orbital energies	1
Population analysis	1

REFERENCE

1. D.T. Clark, *Theor. Chim. Acta* **15**, 225 (1969).

Oxirene C_2H_2O

State	Geometry	Basis set	Energy	Wave function	Reference
$^1A_1$	Experimental	GTO	–150.1090		1

Properties	*Reference*
Dipole moment	1
Orbital energies	1
Population analysis	1

REFERENCE

1. D.T. Clark, *Theor. Chim. Acta* **15**, 225 (1969).

Thi-irene C_2H_2S

State	Geometry	Basis set	Energy	Wave function	Reference
$^1A_1$	Experimental	GTO	–454.8095		1

Properties	*Reference*
Dipole moment	1
Orbital energies	1
Population analysis	1

REFERENCE

1. D.T. Clark, *Theor. Chim. Acta* **15**, 225 (1969).

Oxaziridine CH_3NO

State	Geometry	Basis set	Energy	Wave function	Reference
^1A	Experimental	GTO	–168.6978		1

Properties	*Reference*
Ionization potential	1
Potential surface	1
Population analysis	1
Dipole moment	1

REFERENCE

1. J.M. Lehn, B. Munsch, P. Millie, and A. Veillard, *Theor. Chim. Acta*, **13**, 313 (1969).
   Coordinates given. N inversion barrier given.

Diaziridine CN_2H_4

State	Geometry	Basis set	Energy	Wave function	Reference
^1A	Experimental	GTO	–148.8430		1

Properties	*Reference*
Orbital energies	1
Dipole moment	1
Magnetic properties	1

REFERENCE

1. H. Basch, M.B. Robin, N.A. Kuebler, C. Baker, and D.W. Turner, *J. chem. Phys*. **51**, 52 (1969).

Diazirine CH_2N_2

State	Geometry	Basis set	Energy	Wave function	Reference
$^1A_1$	Experimental	GTO	–146.6980		1
		GTO	–147.7287		3

Properties	*References*
Orbital energies	1, 3
Ionization potential	1
Charge densities	1
Population analysis	1
Dipole moment	1, 3
Quadrupole coupling constant	2, 3
Spectroscopic constants	3

REFERENCES

1. L. Kochanski and J.M. Lehn, *Theor. Chim. Acta* **14**, 281 (1969).
2. L. Kochanski, J.M. Lehn, and B. Levy, *Chem. Phys. Lett.* **4**, 75 (1969).
   GTO basis.
3. M.B. Robin, H. Basch, N.A. Kuebler, K.B. Wiberg, and G.B. Ellison, *J. chem. Phys.* **51**, 45 (1969).

Difluorudiazine $C_2N_2F_2$

State	Geometry	Basis set	Energy	Wave function	Reference
$^1A_1$	Experimental	GTO	−345.3999		2

Properties	*References*
Orbital energies	1, 2
Spectroscopic constants	1, 2
Excited states	1, 2
Population analysis	1
Dipole moment	2

REFERENCES

1. J.R. Lombardi, W. Klemperer, M.B. Robin, H. Basch, and N.A. Kuebler, *J. chem. Phys.* **51**, 33 (1969).
2. M.B. Robin, H. Basch, N.A. Kuebler, K.B. Wiberg, and G.B. Ellison, *J. chem. Phys.* **51**, 45 (1969).

Cyclobutadiene C_4H_4

State	Geometry	Basis set	Energy	Wave function	Reference
^1A_g	Experimental	GTO + CI	−153.4291		1

Properties	*Reference*
Orbital energies	1
Charge density	1

REFERENCE

1. R.J. Buenker and S.D. Peyerimhoff, *J. chem. Phys.* **48**, 354 (1968).

Tetrahedrane C_4H_4

State	Geometry		Basis set	Energy	Wave function	Reference
	R(C–H)	R(C–C)				
$^1A_1$	2.022	2.948	GTO + CI	–153.3710		1

Properties	*Reference*
Orbital energies	1
Ionization potential	1

REFERENCE

1. R.J. Buenker and S.D. Peyerimhoff, *J. Am. chem. Soc.* **91**, 4342 (1969).

Cyclopentane C_5H_{10}

State	Geometry		Basis set	Energy	Wave function	Reference
	R(C–C)	R(C–H)				
^1A	2.89	2.07	GTO	–194.091		1
	Half chair C_2					

REFERENCES

1. J.R. Hoyland, *J. chem. Phys.* **50**, 2775 (1969).
C_s and D_{5h} also considered. Barrier calculated.

2. H. Preuss and R. Janoschek, *J. Molec Struct.* **3**, 423 (1969).
Min. GTO.

Cyclohexane C_6H_{12}

State	Geometry		Basis set	Energy	Wave function	Reference
	R(C–C)	R(C–H)				
$^1A_{1g}$	2.89	2.07	GTO	–232.911		1
	Chair					
	D_{3d}					

REFERENCE

1. J.R. Hoyland, *J. chem. Phys.* **50**, 2775 (1969).
C_{2v}, D_2, and C_2 also considered. Barrier calculated.

2. H. Preuss and R. Janoschek, *J. molec. Struct.* **3**, 423 (1969).
(Min. GTO).

$C_5H_5^-$

State	Geometry	Basis set	Energy	Wave function	Reference
Ground state	Experimental	Min. GTO	–189.42		1, 3

Properties	*Reference*
Orbital energies	1
Ionization potentials	1

REFERENCES

1. H. Preuss and G. Diercksen, *Int. J. Quantum Chem.* **1**, 349 (1967).

2. H. Preuss and G. Diercksen, *Int. J. Quantum Chem.* **1**, 369 (1967).
1s orbital energy discussed.

3. H. Preuss and R. Janoschek, *J. molec. Struct.* **3**, 423 (1969).

LiC_5H_5

REFERENCE

1. H. Preuss and R. Janoschek, *J. molec Struct.* **3**, 423 (1969).
Min. GTO.

Pyrrole C_4H_5N

State	Geometry	Basis set	Energy	Wave function	Reference
$^1A_1$	Experimental	GTO	–207.93135	✓	1

Properties	*Reference*
Orbital energies	1
Population analysis	1
Quadrupole coupling constant	2

REFERENCES

1. E. Clementi, H. Clementi, and D.R. Davis, *J. chem. Phys.* **46**, 4725 (1967).
Coordinates given.
2. L. Kochanski, J.M. Lehn and B. Levy, *Chem. Phys. Lett.* **4**, 75 (1965).
GTO basis.
3. H. Preuss and R. Janoschek, *J. mol. Struct.* **3**, 423 (1969).
Min. GTO.

Furan C_4H_4O

REFERENCE

1. H. Preuss and R. Janoschek, *J. molec. Struct.* **3**, 423 (1969).
Min. GTO.

Oxazole, Iso-oxazole C_3H_4NO

Properties	*References*
Quadrupole coupling constant	1
Ionization potential	2

REFERENCES

1. L. Kochanski, J.M. Lehn, and B. Levy, *Chem. Phys. Lett.* **4**, 75 (1969).
GTO basis.
2. H. Preuss and R. Janoschek, *J. molec. Struct.* **3**, 423 (1969).
Min. GTO gives $E = -240.7$.

Imidazole, Pyrazole $C_3N_2H_5$

Properties	*Reference*
Quadrupole coupling constant	1

REFERENCE

1. L. Kochanski, J.M. Lehn, and B. Levy, *Chem. Phys. Lett.* **4**, 75 (1969).
GTO basis.

Benzene C_6H_6

State	Geometry		Basis set	Energy	Wave function	Reference
	R(C–H)	R(C–C)				
$^1A_{1g}$			GTO	–227.27		3, 8
	2.63	2.17	GTO	–227.304		5
			Gaussian lobe + CI	–230.3745		6
	2.637	2.051	GTO	–230.463		2

Properties	*References*
Orbital energies	1, 2, 3, 6, 7
Excitation energies	1, 2
Ionization potentials	1 2, 3, 7
Population analysis	2
Spectroscopic constants	5
Potential curve	5

REFERENCES

1. J.M. Schulman and J.W. Moskowitz, *J. chem. Phys.* **43**, 3287 (1965). GTO basis. Excited states $^3B_{1u}$, $^3B_{2u}$, $^3E_{1u}$, $^1B_{1u}$, $^1B_{2u}$, $^1E_{1u}$ by virtual orbitals.

2. J.M. Schulman and J.W. Moskowitz, *J. chem. Phys.* **47**, 3491 (1967). Ions are also calculated.

3. G. Diercksen and H. Preuss, *Int. J. Quantum Chem* **1**, 357 (1967).

4. H. Preuss and G. Diercksen, *Int. J. Quantum Chem.* **1**, 369 (1967). 1s orbital energies discussed.

5. R. Janoschek, H. Preuss, and G. Diercksen, *Int. J. Quantum Chem.* **1S**, 209 (1967).

6. R.J. Buenker, J.L. Whitten, and J.D. Petke, *J. chem. Phys.* **49**, 2261 (1968).

7. B.O. Jonsson and E. Lundholm, *Ark. Fys.* **39**, 65 (1968). 1s neglected.

8. H. Preuss and R. Janoschek, *J. molec. Struct.* **3**, 423 (1969). Min. GTO.

$C_6H_7^+$

REFERENCE

1. H. Preuss and R. Janoschek, *J. molec. Struct.* **3**, 423 (1969). Min. GTO. (protonated benzene).

Li C_6H_5

State	Geometry	Basis set	Energy	Wave function	Reference
Ground state	Experimental	GTO	–196.231		1

Properties	*Reference*
Potential surface (Li-C stretch)	1
Ionization potential	1
Spectroscopic constants	1

REFERENCE

1. R. Janoschek, G. Diercksen, and H. Preuss, *Int. J. Quantum Chem.* **1S**, 265 (1967).

Pyridine C_5H_5N

State	Geometry	Basis set	Energy	Wave function	Reference
$^1A_1$	Experimental	GTO	–245.62194	✓	1
		GTO	–246.32653		2

Properties	*References*
Orbital energies	1, 2

Properties	*References*
Population analysis	1
Dipole moment	2
Quadrupole coupling constant	3

REFERENCES

1. E. Clementi, *J. chem. Phys.* **46**, 4731 (1967). Coordinates given.

2. J.D. Petke, J.L. Whitten, and J.A. Ryan, *J. chem. Phys.* **48**, 953 (1968).

3. L. Kochanski, J.M. Lehn, and B. Levy, *Chem. Phys. Lett.* **4**, 75 (1969). GTO basis.

Pyridine cation $C_5H_5N^-$

State	Geometry	Basis set	Energy	Wave function	Reference
Ground state	Experimental	GTO	–245.19708	✓	1

Properties	*Reference*
Population analysis	1
Ionization potential	1

REFERENCE

1. E. Clementi, *J. chem. Phys.* **47**, 4485 (1967). Geometry as in pyridine (ref.1).

Pyrazine $C_4H_4N_2$

State	Geometry	Basis set	Energy	Wave function	Reference
^1A_g	Experimental	GTO	–261.5542	✓	1
		GTO	–262.25466		2

Properties	*References*
Orbital energies	1, 2
Population analysis	1
Dipole moment	2
Quadrupole coupling constant	3

REFERENCES

1. E. Clementi, *J. chem. Phys.* **46**, 4737 (1967).
Coordinates given.

2. J.D. Petke, J.L. Whitten, and J.A. Ryan, *J. chem. Phys.* **48**, 953 (1968).

3. L. Kochanski, J.M. Lehn, and B. Levy, *Chem. Phys. Lett.* **4**, 75 (1965).
GTO basis.

C_7H_7 (Benzyl)

REFERENCE

1. H. Preuss and R. Janoschek, *J. molec. Struct.* **3**, 423 (1969).
Min. GTO.

C_8H_8 (Cubane)

REFERENCE

1. H. Preuss and R. Janoschek, *J. molec. Struct.* **3**, 423 (1969).

Naphthalene $C_{10}H_8$

State	Geometry	Basis set	Energy	Wave function	Reference
${}^1A_g$	Experimental	GTO	–382.7883		2

Properties	*References*
Ionization potential	1, 2
Charge densities	2
Orbital energies	2

REFERENCES

1. H. Preuss, *Int. J. Quantum Chem.* **2**, 651 (1968).
GTO basis.

2. R.J. Buenker and S.D. Peyerimhoff, *Chem. Phys. Lett.* **3**, 37 (1969).
Excited states.

Azulene $C_{10}H_8$

State	Geometry	Basis set	Energy	Wave function	Reference
$^1A_1$	Experimental	GTO	–382.7082		1

Properties	*Reference*
Ionization potential	1
Charge densities	1
Orbital energies	1

REFERENCE

1. R.J. Buenker and S.D. Peyerimhoff, *Chem. Phys. Lett.* **3**, 37 (1969).
Excited states.

α– Tocopherol

REFERENCE

1. R. Janoschek, *Int. J. Quantum Chem.* **2**, 707 (1968).
Oxygen radical $\rightleftharpoons$ methylene radical reaction surface calculated as part of this molecule. GTO basis.